DESIGN FOR MAINTAINABILITY

Benchmarks for Quality Buildings

DESIGN FOR MAINTAINABILITY

Benchmarks for Quality Buildings

CHEW Yit Lin, Michael
Ashan ASMONE
Sheila CONEJOS

Department of Building, National University of Singapore

World Scientific

NEW JERSEY • LONDON • SINGAPORE • BEIJING • SHANGHAI • HONG KONG • TAIPEI • CHENNAI • TOKYO

Published by

World Scientific Publishing Co. Pte. Ltd.
5 Toh Tuck Link, Singapore 596224
USA office: 27 Warren Street, Suite 401-402, Hackensack, NJ 07601
UK office: 57 Shelton Street, Covent Garden, London WC2H 9HE

National Library Board, Singapore Cataloguing in Publication Data
Name(s): Chew, M. Y. L. | Asmone, Ashan, author. | Conejos, Sheila, author.
Title: Design for maintainability : benchmarks for quality buildings /
Chew Yit Lin Michael, Ashan Asmone, Sheila Conejos.
Description: Singapore : World Scientific Publishing Co. Pte. Ltd., [2018] |
Includes bibliographical references and index.
Identifier(s): OCN 1004040082 | ISBN 978-981-3230-59-0 (hardcover) |
ISBN 978-981-3232-95-2 (paperback)
Subject(s): LCSH: Building. | Buildings--Maintenance. | Standards, Engineering.
Classification: DDC 690--dc23

British Library Cataloguing-in-Publication Data
A catalogue record for this book is available from the British Library.

For any available supplementary material, please visit
http://www.worldscientific.com/worldscibooks/10.1142/10722#t=suppl

Desk Editor: Amanda Yun

Typeset by Stallion Press
Email: enquiries@stallionpress.com

Preface

Responding to calls from developers and building professionals, *Design for Maintainability* is written as a guide to aid in achieving optimum performance throughout the lifespan of a facility within the minimum life cycle cost.

With the motto of "doing it right the first time", this book focuses on spearheading the integration of designers, constructors and facility managers (FM), on the outset of the planning/design stage, by providing easy to read tables summarising (1) knowledge learnt from past mistakes and (2) maintainability benchmarks, to ensure high maintainability. Based on the predictive/preventive approach, the tables serve to define acceptable standards in design, construction and operation practices. The text covers technical issues related to major components of a facility for basements, wet areas, façades, common areas, roofs and M&E (mechanical and electrical) systems.

Readers who seek further information beyond the tables of this reference book may refer to the associated textbook ***Maintainability of Facilities: Green FM for Building Professionals***(2nd Edition). The textbook also provides access for an online "Maintainability of Building" database (http://www. hpbc.bdg.nus.edu.sg).

Thank you.

Chew Yit Lin, Michael
Professor, Department of Building
School of Design and Environment
National University of Singapore

Disclaimer

The data provided in this publication is prepared in good faith. Nevertheless, no statements or guarantees are articulated or indicated in relating the accuracy or completeness of this publication in regards to any specific user's requirements. Information in this publication is specified with the condition that its applicability be considered in terms of the specific project's situations. Designers and allied professionals should depend on their own knowledge and capability in appraising and utilising any information explained herein. In using such information, they should be careful of their own safety and the safety of others, taking into account that the use and reliance on the information contained herein will be borne by the user. The authors will not be held responsible for any damage, loss or expense resulting from the use of the information stipulated in this publication.

Contents

Introduction

The investment on building maintenance represents almost half of the total turnover of the construction industry and such reduction of resources has a direct effect on a nation's economy. Recent studies have shown that the main factors that lead to building operations and maintenance problems are due to faulty designs, faulty construction, financial factors and maintenance related defects. The lack of maintainability considerations during the design and construction stages lead to building defects which account for expenditures of billions of dollars throughout the building's lifecycle. More so, the potentially unsafe conditions of buildings (specifically on high rise building façades) can be detrimental to the lives and health of construction workers, and may even jeopardise public safety and surrounding properties if left unaddressed. Thus, to ensure the maintainability of buildings, the benchmark for quality must be established from the design stage itself in order to achieve quality design and construction, and maintenance best practices.

How to Use this Book

The book is presented in six chapters highlighting the major building components: (1) basements, (2) wet areas, (3) façades, (4) roofs, (5) common areas and (6) M&E systems. Each chapter addresses the common defects and the corresponding design/construction/FM issues, with standards/guidelines/recommendations for the structural, architectural and service components of the building component, to be taken into account at the outset of the planning/design stage. Although the chapters are organised according to the normal physical order of a building (starting with the basement, followed by the wet areas, façades, roof, then common areas and M&E),

these chapters may be utilised individually to deal with specific problems, or may be combined for wide-ranging projects.

For example, the 'corrosion and spalling of concrete' in a basement is classified as a structural defect, highlighted with photographs taken from case studies, and the corresponding design, construction, and maintenance standards/guidelines to be considered at the outset of planning/design stage.

Structural

Problem	Design	Construction	Maintenance
Corrosion/spalling of concrete 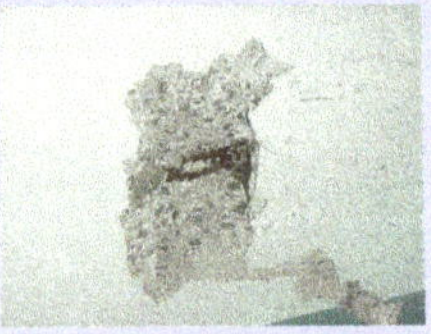 Spalled concrete and corrosion of exposed rebars  Extreme spalling and corrosion of basement	Comply with the design of concrete as per SS EN 1992-1-1:2008, SS EN 1992-1-2:2008 (2015), SS CP 65-1:1999. Specify admixtures (e.g. water reducing agent, pozzolanic products, pore refiner, etc.) to reduce permeability. Alternatively, for corrosion control in special areas with a high-risk of water penetration, an electrochemical treatment can be specified; a process where electrochemical drying of concrete occurs by passing a current through the reinforcement, similar in principle to cathodic protection (BS EN 1504-9:2008).	For the construction of basements, refer to the guidelines and provisions stipulated in SS CP 4:2003 (2012) (see also BS 8004:2015). Maintain water–cement ratio and the required aggregate grading during basement construction.[1] Verify aggregate quality in compliance with ASTM C33/C33M-16e1. Use corrosion resistant bars and corrosion inhibitors. Apply proper vibration (compaction) and curing. Consider concrete sealing to avoid exposing the aggregate in concrete work (mitigate pitting, scaling, spalling, powdering, or chalking of concrete).	Conduct regular inspection in accordance to BS 8210:2012 to identify defects and ensure repair work is carried out before associated damage can occur. Testing of concrete via depth of carbonation to identify possible corrosion of rebar as per BS 1881-210:2013; or via phenolphthalein method as per BS EN 14630:2006. Repair and protect damaged concrete due to corrosion of reinforced steel as per BS EN 1504-9:2008.

The standards and guidelines specified should only be used as a reference. Supplementary sources and references are provided for further reading. A comprehensive index is provided to further assist the reader in locating a particular topic of interest within the book.

Chapter 1

Basements

Introduction

A basement refers to any floor level below the first story in a building that is partly or completely below the exterior grade plane [1] with at least one-half of its floor-to-ceiling height above curb level or above the average natural grade [2]. The space can be utilised in the form of carparks, plant rooms and workshops, or storage, residential, retail and or office areas. As an underground structure, it is susceptible to ground water ingress and external hydrostatic pressure, and subject to a rapid degree of deterioration compared to other building components since it is directly exposed to the surrounding soil conditions. The performance requirements for water-tightness varies depending on the basement usage. For example: (a) some seepage and damp patches are tolerable for carpark, plant rooms and workshops; (b) no water penetration but moisture vapour is tolerable for workshops and plant rooms requiring drier environments (e.g. retail storage areas); (c) dry environment should be provided for ventilated residential and working areas (e.g. offices, restaurants, etc.); and (d) a totally dry environment is necessary for archives and stores requiring controlled environments [3]. Durability is important in the design, construction and maintenance of basements — on top of keeping it moist and damp free — to lessen the frequency of repair and maintenance. It is essential to reduce hydrostatic pressure around the basement structure and prevent the entry of stormwater. The use of reinforced concrete for the basement's outer wall and the feasible application of a waterproofing system is recommended to prevent water

Basement area of a retail commercial complex.

seepage and ensure that the membrane does not lose its protective capacity [4, 5]. The waterproofing membrane material should be selected based on a few criteria, including (a) long term water-tightness; (b) durability in exposure; (c) ease of application; (d) curing period; (e) flexibility; (f) tensile property; (g) puncture resistance; and (h) compatibility with substrate [5, 6].

The utmost major problem in basements is the failure to attain water-tightness, though water seepage defects are common basement problems. Defects usually surface at a later stage due to changes in environmental conditions such as rising ground water, sump effect of excavation, differential settlement, etc. Basements also suffer from poor design, construction or maintenance of non-basement ancillary elements, such as a planter box or swimming pool at plaza level. The correct design, construction and maintenance practices for basements will help inhibit prevalent defects while ensuring efficient buildings with high maintainability. Important guidelines for the design, construction and maintenance of basements are highlighted in the succeeding tables, to address each corresponding basement defect. The identified structural defects refer to corrosion/spalling of concrete, seepage through porous concrete and cracks, and flood control. The identified architectural defects are concerned with construction joint installation, seepage through joints, water ponding and waterproofing issues, wall and flooring finishes. The identified service defects pertain to drainage and service penetration issues. Fire protection guidelines are also provided as a relevant requirement for basement areas.

The water-tightness of a basement depends mostly on the waterproofing design detailing which includes the waterproofing's capability over penetrations, projections and joints. Drainage design detailing is critical to inhibit water build-up behind basement walls and subsequent water seepage/leakage. Careful consideration is necessary to certify that the selected material is appropriate for the environment of the basement, and that it matches the other materials used, so as not to compromise the overall lifespan of the basement due to untimely failures of mismatched materials. Quality control during construction is also a critical issue [7] especially for basements. Basement construction requires special attention due to its exposure to hydrostatic pressure. Provision of waterstops and drainage details are common devices in managing seepage through a basement's construction joints. Good workmanship and the thorough protection of an installed waterproofing membrane can augment the waterproofing's useful life to a substantial extent. Proper curing of the waterproofing is also crucial to preventing excessive shrinkage and cracking which will eventually lead to water seepage. The major critical elements (e.g. RCC structure, positive waterproofing or waterstops) are often beyond regular inspection and maintenance. Defects pertaining to these elements demonstrate visible signs on internal finishes, thus the locating of any early signs of leakage or dampness during post occupancy period must be performed. The detrimental effect of these external factors (e.g. mild to extreme, based on hydrostatic pressure, concentration of sulphur or chloride, acidity, contaminant, gas, radioactive material, etc.) should be taken into consideration in the maintenance stage [8].

1.1 Structural

The structural components of a building (e.g. foundation, floor, slab, walls, and other load bearing and non-load bearing components) require design, construction and maintenance specifications to prevent premature deterioration and to help maximise the basement's performance. The basement's design should adhere to prevailing industry standards relating to foundation construction, structural stability, fire resistance, etc., whereas the builders must abide by local regulations and design drawings and specifications to ensure good quality in construction.

Problem	Design	Construction	Maintenance
Corrosion/spalling of concrete  Spalled concrete and corrosion of exposed rebars  Extreme spalling and corrosion of basement	Comply with the design of concrete as per SS EN 1992-1-1:2008, SS EN 1992-1-2:2008 (2015), SS CP 65-1:1999. Specify admixtures (e.g. water reducing agent, pozzolanic products, pore refiner, etc.) to reduce permeability. Alternatively, for corrosion control in special areas with a high-risk of water penetration, an electrochemical treatment can be specified; a process where electrochemical drying of concrete occurs by passing a current through the reinforcement, similar in principle to cathodic protection (BS EN 1504-9:2008).	For the construction of basements, refer to the guidelines and provisions stipulated in SS CP 4:2003 (2012) (see also BS 8004:2015). Maintain water–cement ratio and the required aggregate grading during basement construction.[1] Verify aggregate quality in compliance with ASTM C33/C33M-16e1. Use corrosion resistant bars and corrosion inhibitors. Apply proper vibration (compaction) and curing. Consider concrete sealing to avoid exposing the aggregate in concrete work (mitigate pitting, scaling, spalling, powdering, or chalking of concrete).	Conduct regular inspection in accordance to BS 8210:2012 to identify defects and ensure repair work is carried out before associated damage can occur. Testing of concrete via depth of carbonation to identify possible corrosion of rebar as per BS 1881-210:2013; or via phenolphthalein method as per BS EN 14630:2006. Repair and protect damaged concrete due to corrosion of reinforced steel as per BS EN 1504-9:2008.
Seepage through cracks  Water stains and efflorescence along crack lines	Surface cracks (caused by applied loads, or thermal and shrinkage effects) should not exceed width of 0.3mm (SS EN 1992-1-1:2008). Provide rebar layout to facilitate vibrations/ compactions.	Installation of waterproofing as per BS 8102:2009 (see also, SS CP 82:1999). Careful supervision of all stages of construction. Construct for waterproofing with due understanding of site conditions and construction method as per BS 8102:2009.	Conduct routine inspection of basement areas for visible signs of water seepage such as efflorescence, especially at high risk areas such as joints and pipe penetrations.

[1] *See Maintainability of Buildings, Basement.* Available at: http://www.hpbc.bdg.nus.edu.sg/?page_id=88&page=2 (accessed May 12, 2017) for more information on grading of structural concrete, waterproofing system, drainage, flooring and wall finishes for basements.

Problem	Design	Construction	Maintenance
 Water ponding and subsequent rising dampness on the wall 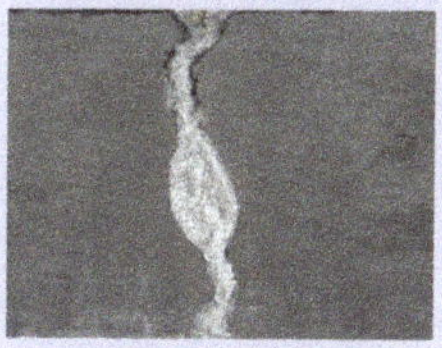 Water seepage and efflorescence along crack lines	Water-resistance is a primary design concern (BS EN 1992-3: 2006) — seepage through cracks in structural elements is unacceptable for basements. Provide waterproofing with reinforced concrete design as per BS EN 1992-1-1:2004+ A1:2014 and BS EN 1992-3:2006 for crack prohibition at stress concentrated areas (e.g. joints, interfaces).	Special considerations on stress concentrated areas such as joints and interfaces between different materials. A crack width limit of 0.2 mm normally applies to all cracks, irrespective of whether or not they pass completely through the section. Where the appearance of a structure is considered to be aesthetically critical, a limit of 0.1 mm is recommended (BS EN 1992-3:2006).	Protect against ingress and increase moisture resistance with (e.g.) hydrophobic impregnation, coating (BS EN 1504-2:2004), or filling of cracks (BS EN 1504-5:2013). Concrete restoration with (e.g.) hand-applied mortar, recasting or spraying with concrete/ mortar (BS EN 1504-3: 2005). Restoring passivity by (e.g.) increasing cover with additional mortar/concrete, replace contaminated or carbonated concrete (BS EN 1504-3:2005).
Seepage through porous concrete 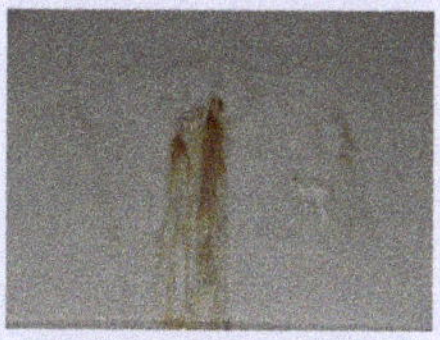 Efflorescence and water stains due to seepage	Monolithic basement construction (wall/ floor) is required for water-tightness. Specify the proper waterproofing with suitable material and consider aggressive conditions and proper mix design as per BS 8102: 2009 (see also SS CP 82:1999).	For quality workmanship, adhere to the recommended curing process; ensure sufficient reinforcement cover; and proper vibration (for well compacted mortar) of concrete to avoid honeycombs as per CONQUAS 21 recommendations [9].	Regular inspections of basement area as per BS 8210:2012. Repair identified seepage defects by injecting cracks, voids or interstices (refer to EN 1504-5:2013). Enforce a full recorded survey of condition before problems are hidden below patch repairs, coatings or waterproofing (BS EN 1504-9:2008).

(*Continued*)

(*Continued*)

Problem	Design	Construction	Maintenance
Flood Control  Flooded mall underpass  Flooded basement parking	For basements in water-bearing soils, ensure that the upward water pressure is uniform below the whole area of the floor. It must be capable of resisting the total pressure less the weight of the floor. The walls must be designed to resist the horizontal pressures due to the waterlogged ground, and prevent the basement from flooding (BS 8004:2015).	During construction, ensure that there is always an excess of downward load to exceed the worst credible upward force due to the water by a substantial margin (BS 8004:2015). Permanent deep basement structures create permanent change in ground water patterns when its dead load exceeds the total upward force on the structure.	Check and confirm by inspection and testing for any deviations from as-built drawings; any indications of defective or substandard construction; indications of severe local environments from ponding, waterproofing breakdown, and seepage; and for the current trends of deterioration and likely long-term trends (BS EN 1504-9:2008).
 Flood Barrier (left: not activated, right: activated) [10]	Provide design features to protect basements from flash floods (e.g. humps and flood barriers) [10]. Provide alternative surface water retention systems; and suggest preparation of standby sand bags and floorboards for residential buildings located in low-lying areas [11, 12].	During construction, the level of groundwater near the basement needs to be controlled by pumping or other means. Consider methods such as open pumping, pre-drainage, cut-offs or exclusion. Water that has infiltrated the building due to structural porosity needs to be drained using dedicated pipes to sumps and subsequent pumping [3].	Conduct periodic inspection for ponding and make visual observations of free flowing water towards the outlets to ensure that water has drained off, so as to avoid leftover ponding on reinforced concrete (BS 8221-1: 2012, SS 509-1:2015). Monitor basement carpark water-level using sensors and warn building users in a timely manner [10].

BASEMENT WATERPROOFING: TYPES AND DURABILITY

Description Basement Waterproofing	Expected Lifespan (Years)	Adjustment Factors
Preformed membrane Sheet/roll applied waterproofing with base material to BS 8102:2009, SS CP 82:1999. Methods of fixing include self-adhesives, adhesive bonding, chemical welding, hot air welding, and mechanical fixing, in accordance to manufacturer's instruction. • Bituminous: asphalt, bitumen, modified bitumen • Non Bituminous: Ethylene-Cpolymer Bitumen (ECB), Polyvinyl Chloride (PVC), Polyethylene (PE), Chlorinated Polyethylene (CPE), Butyl Rubber, Ethylene Propylene Diene (EPDM), Neoprene, Polyisobutylene (PIB), Bentonite clays	10	• Existence of aggressive ground water*: –15% (years). • Hydrostatic head** exceeding system design (except crystalline system): –10% (years). • Test certificate from approved local testing authority: +15% (years).
Liquid applied system Liquid (cold/hot) applied with base material to BS 8102:2009, SS CP 82:1999. Methods of fixing include by means of a squeegee, roller, brush, towel or spray, in accordance to manufacturer's instruction. • Bituminous • Non Bituminous: acrylic, cementitious, epoxy resin, hypalon, neoprene, polyurethane, polysulphide, rubber and silicone Liquid (cold/hot) applied with the base material not in accordance to CP 82:1999. Methods of fixing include by means of a squeegee, roller, brush, trowel or spray, in accordance to manufacturer's instruction.	10 <10	
Integral system* Integral waterproofing system. Methods of application include by means of a masonry brush or compressed air-spray equipment, or by dry shake or as an admixture in accordance to manufacturer's instruction. • crystalline waterproofing • chemical admixture • waterproofing screed	10+	

**See Maintainability of Buildings, Waterproofing.* Available at: http://www.hpbc.bdg.nus.edu.sg/?page_id=384&page=2 (Accessed May 12, 2017)for more on material information (including material properties, defect identification, cleaning and repair) of waterproofing systems.

1.2 Architectural

Architectural components define the character of the building, which include the finishes, furnishings, and other elements that contribute to the building's aesthetic value and liveability. The design, construction and maintenance specifications for a basement's architectural components will ensure optimal performance and better quality workmanship of the building.

Problem	Design	Construction	Maintenance
Proper installation of construction joints 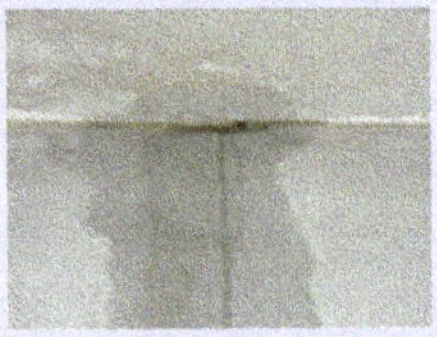 Water seepage through joint 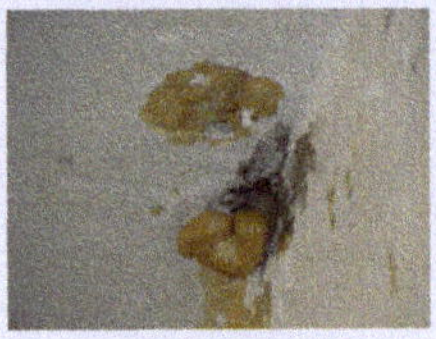 Chemical injection at a wall-to-ceiling construction joint  Cracks along joints, and joint sealants need replacing	Specify the use of expansion joints where effects of temperature and moisture are too large to absorb as a strain in order to separate framed structures from joints' sections (BS 6093:2006+ A1:2013). Refer to the selection and specification of sealants as per BS 6213:2000+A1:2010. Refer to BS EN ISO 11600: 2003+A1:2011 for a systematic performance classification scheme for sealants applied in building and construction. Provide internal/external waterstops to accommodate differential movements (Flanges of water-stops tied to RF in adequate intervals) [7].	Seal construction joints to withstand multi-directional stresses. Ensure proper installation of waterstops [7]. Provide proper surface preparation for the installation of sealants and gaskets (ensure that gaskets should not be stretched during installation) (BS 6093: 2006+A1:2013). Slow down the drying of concrete to avoid plastic shrinkage and provide joints/planes of weakness to confine cracking to determined positions (BS 6093:2006+A1: 2013). Joint gap width can be changed from the time of erection to sealing. Hence, the best time for sealant application in movement joints is when joint gap is at the mean trending to the maximum (BS 6093: 2006+A1:2013).	Conduct visual inspection on the integrity of the joints (BS 8102:2009, SS CP 82:1999). Use non-destructive field diagnostic techniques to identify seepage through joints where visual inspection is not sufficient (e.g. Infra-red thermography for moisture detection, portable microwave tomography to identify source of leakage).[2] Replace old sealant with a suitable one (ASTM C1193-16). Joints inspected at intervals equivalent to one-fifth of their expected life (additio-nally, all joints subject to movement be inspected for signs of premature failure after first year in service) (BS 6093: 2006+A1: 2013)

(*Continued*)

[2] See *Maintainability of Buildings, Non-Destructive Testing*. Available at: http://www.hpbc.bdg.nus.edu.sg/?page_id=31 (accessed May 12, 2017) for more information on the field diagnostics techniques used to identify basement water seepage. It includes an exhaustive list of non-destructive tests used to detect internal, surface or concealed defects in various building systems.

(*Continued*)

Problem	Design	Construction	Maintenance
Seepage through joints  Algae and water stains at joints  Efflorescence, water stains and mould growth at joints	Specify proper sealant design at joints to prevent seepage caused by differential movements (BS EN 15651-3:2017, SS CP 82:1999). Water penetration is prevented by incorporating sealants, sealing strips, gaskets and baffles (BS 8102: 2009). Provide comprehensive design detailing of joints to avoid the promotion of mould and plant growth; discolouration due to UV radiation and biological, physical or chemical action; showing of internal structure (in part or all); and dust collection (BS 6093:2006+A1: 2013).	The use of wet joints is essential in minimising water seepage through joint areas. Ensure its proper installation with suitable backer rod and sealant application [13]. For vertical joints, a baffle provides the overlapping needed to stop water seepage. The baffle needs to extend beyond the groove and not shortened due to creep (BS 6093:2006+ A1:2013). Follow methods for cementitious waterproofing as per BS EN 1992-1-1:2004+A1:2014 and refer to BS EN 1992-3:2006 for jointing of membranes.	Conduct routine visual inspections of joints and look for tell-tale signs of water seepage in basement areas (e.g. deteriorated or damaged wall/ floor/ceiling material, biological growth[3]). If seepage through joints is suspected, non-destructive tests can assist with the identification of the sources of water. A water spraying test should be carried out at the precast joints to check for water-tightness (BS 8102:2009, SS CP 82:1999).

[3] See *Maintainability of Buildings, Fungus*. Available at: http://www.hpbc.bdg.nus.edu.sg/?page_id=266 (accessed May 12, 2017) for more information on cases of biological growth in buildings and its remediation.

(*Continued*)

Problem	Design	Construction	Maintenance
Water ponding  Water ponding in a basement	Conform to the proper drainage design (slope/ outlet) [13;14]. Comply with the proper compacting of ground/ soil to maintain even settlement (BS 8004:2015, SS CP 4:2003(2012)). Abide to requirements in basement construction as per PUB: Code of Practice on Surface Water Drainage [11].	Perform proper levelling of floor surface to avoid ponding in falls for wet area [14]. Establish full specification and procedure for repair, propping and testing, and handing over with the as-built drawings [15].	Conduct periodic inspection for ponding and make visual observations of free flowing water towards the outlets. Ensure that water has drained off to avoid leftover ponding on the reinforced concrete. Drains should be inspected for efficiency prior to water washing (BS 8221-1:2012, SS 509-1:2015).
Waterproofing  Seepage through slab and wall due to poor waterproofing  Water seepage due to poor waterproofing	Minimise wall joints in wet areas. Kerbs to be constructed at the base of walls to halt lateral movement of water. Basement masonry structure is not recommended. Select waterproofing types A, B, or C* or a combination of them, depending on the performance requirements of usage cases (BS 8102:2009, SS CP 82:1999). Conform to the use of damp-proof courses as per BS 6398:1983, BS 8215:1991.	Ensure proper surface preparation of concrete using mechanical or chemical cleaning. Ensure that the substrate is dry, so as to avoid blistering of non-breathing waterproofing material. Allow for good ventilation and check with a moisture meter. Conversely, pre-saturate substrate if cementitious membrane systems are used for improved adhesion.	Concrete surface cleaning and repair applications of a continuous waterproofing barrier system should be carried out (BS 8102:2009, SS CP 82:1999). The cleaning method selected should be chosen on the basis of minimising damage to existing surfaces. Aggregate rich surfaces should not be cleaned by processes that can damage the binder (BS 8221-1:2012, SS 509-1:2015).

(*Continued*)

(*Continued*)

Problem	Design	Construction	Maintenance
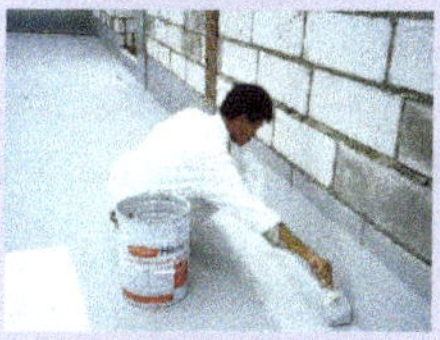 Example of Type A* waterproofing system  Example of Type B* waterproofing system  Example of Type C* waterproofing system	Architectural, structural, and M&E coordination drawings of affected wet areas are reviewed together to ensure proper integration and reliability of waterproofing system. Provide proper design for adequate drop of cast concrete to maintain floor gradient and to prevent water from migrating to the dry area. A screed should be laid over the membrane as a protective measure and sloped towards the floor outlet [9]. Protect basement joints by using waterstops of at least 240mm width. Waterstop design should be compatible with the waterproofing membrane and the joint type.	Use of continuous angle fillet (25×25 mm, 1:3 cement sand with 1:4 bonding agent and water) for membrane to lay gradually over internal corners and floor–wall joints. Waterproofing products mixed as per manufacturer's specifications in a controlled environment. Application also as per manufacturer's specifications [9]. Protective measures should be taken into consideration to prevent the waterproofing membrane from damage by any activities on site (e.g. by use of Styrofoam) (BS 8102: 2009, SS CP 82:1999).	Ensure that water collecting points, channels and scupper drains are cleared, cleaned, free from debris and mortar droppings and adequately ventilated to prevent any build-up of saturated atmosphere inside the cavity (BS 8102:2009, SS CP 82:1999). Take precautions to avoid leaks. Any joints suspected or confirmed to be leaking should receive immediate attention as delay can cause extensive damage. Proper insulation should be provided for all external and exposed piping (BS 8210:2012).

***CLASSIFICATION OF WATERPROOFING SYSTEMS [8]**

Type A (Tanked Protection) — the structure itself does not prevent water ingress. Protection is dependent on a total water or water and vapour barrier system applied internally or externally.

Type B (Structurally Integral Protection) — refers to admixtures that are supposed to decrease the water permeability of structural concrete. This is not just a matter of concrete mix but also involves issues in structural design and the manner of handling of the concrete in the field. It can be applied by brush (as a coating), dry shake, or as an admixture.

Type C (Drained Protection) — the drained cavity wall and floor construction provides a high level of safeguard. Provision of a ventilated cavity and horizontal damp-proof membrane prevents moisture ingress.

Problem	Design	Construction	Maintenance
Wall finishes — Paint   Blistering and efflorescence 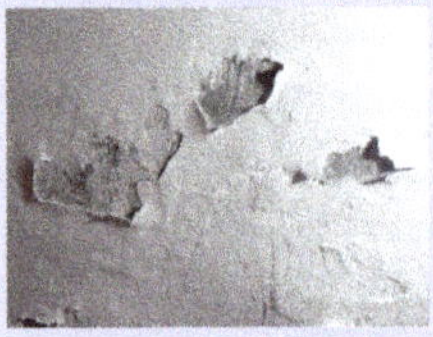 Peeling and flaking of paint  Paint peeling on wall	Specify the use of a breathable paint system to reduce trapped moisture and to avoid wetness and dampness of basement walls. Fungi-resistant paint is also required. Avoid the use of Alkyd based paint on concrete surfaces that may lead to saponification (i.e. the formation of oily patches) (BS EN 1504-2:2004). Thoroughly investigate ground conditions prior to selecting the waterproofing system. The design of the waterproofing system's detailing is also of paramount importance to protecting the basement. Protect basement construction joints using waterstops which are compatible with the waterproofing membrane. Waterstop design[5] should be suitable for the joint (BS 8102:2009, SS CP 82:1999).	Ensure adequate curing of substrate before paint application to avoid shrinkage cracks. Moisture content should not exceed 6%. Clean surface and use clean tools. Avoid prolonged storing, inadequate stirring, use of incompatible thinner/ solvent, or mixing with leftover paints from previous batches. (BS 6150:2006+A1: 2014, SS 542:2008). Ensure that coatings are always applied in a minimum of two coats. The material should possess some degree of flexibility (i.e. be elastomeric) to reduce the risk of cracking due to thermal/moisture movements (BS 8000-0:2014, SS 150:2015). If it is brush-applied, the second coat should be at right-angles to the first coat to help eliminate pinholes and avoid chemical attacks (BS 8000-0:2014, SS 150:2015).	Inspect at reasonable intervals to identify necessary repairs (BS 6150: 2006+ A1:2014, SS 542:2008) (depending on type of coating, degree of exposure to elements, and accessibility, it can range between 3–10 years). Walls that are lightly-soiled should be washed with water and a mild detergent. For severe soiling, wash with a strong alkali solution in warm water. Re-paint thee walls regularly. Repair hairline cracks by re-painting the wall with a flexible sealant/elastomeric paint to seal the cracks. For wider cracks, engage a contractor to carry out repair works.[4] Paint defects can be easily identified during visual inspections. In recurrent cases, use non-destructive testing (e.g. moisture meter) to assess surface dampness from structural dampness and identify water leakages.

(Continued)

[4] See *Maintainability of Buildings, Paint.* Available at: http://www.hpbc.bdg.nus.edu.sg/?page_id=385(accessed May 12, 2017) for material properties and durability data on different types of paints. Further information on cleaning, maintenance and repairing defects is referred therein.

[5] These can be installed either internally or externally, and external waterstops should be at least 240mm in width in order to be effective.

(Continued)

Problem	Design	Construction	Maintenance
Flooring  Damp/dirt stain 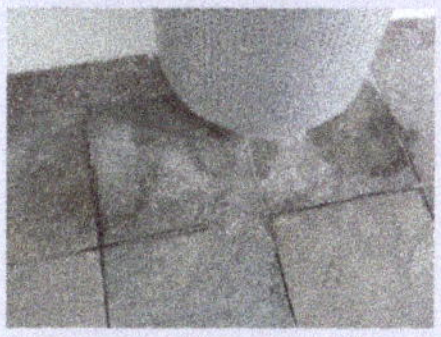 Water mark and moulds along column joint 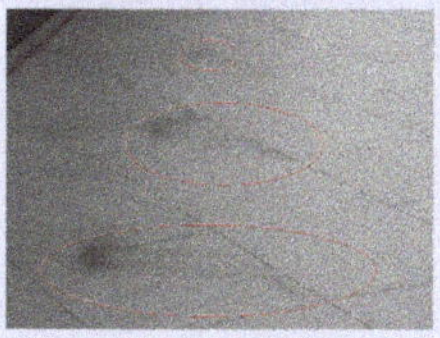 Floor stains	Ensure the design works as a monolithic unit to comply with requirements for durability, strength, dimensional stability, resistance to wear, slip and dampness (BS 8204-2:2003+A2:2011). Specify type of additives to use (e.g. air-entraining superplastisizing, set retarding admixtures, low heat Portland cement, in-surface organic resin sealer). Avoid Alkyd based paints that may cause saponification [7]. Specify suitable, good-quality aggregate for concrete (SS EN 12620:2008) and mortar (BS EN 13139:2002).	Provide falls (max 1:40) on base concrete for good drainage. Fall should be in finishing not in screed. Insert induced contraction joints at about 6 m to 8 m from centre-to-centre (c.t.c.) to avoid cracking. It should be sawn within a day of laying and should be in a proper straight line to avoid differential contraction [16]. Checks for levelness and surface regularities should be done within 24 hours of casting, as delays can mean no possible rectifications. Measure from the underside of a 2 m straightedge (between points of contact) placed anywhere on surface and using a slip gauge. (BS 8204-1: 2003+A1:2009).	Clean floors regularly and perform daily sweeping of floors. Perform weekly washing of floors and monthly scrubbing of floors.[6] Identify and clean grease stains using an aqueous solution of alkaline salts (caustic soda, sodium meta-silicate, tri-sodium phosphate, appropriate proprietary detergent, etc.). Identify sealant deterioration in joints through routine inspection. Clean joints and reseal with hard sealants (e.g. synthetic resin composite). Movement joints need flexible sealants [7].

[6]See *Maintainability of Buildings, Basement Floor Screed.* Available at: http://www.hpbc.bdg.nus.edu.sg/?page_id=383 (accessed May 12, 2017) for further information on common defects in basement floors and their respective repair methods.

1.3 Services

Service components refer to the vertical and horizontal circulation systems, electro-mechanical and sanitary connections in buildings. Each issue is presented with its corresponding design, construction and maintenance measures to ensure better performance.

Problem	Design	Construction	Maintenance
Drainage  Provision of drainage box to channel away ingressed water	Provide an internal basement drainage design since it is more cost-effective and affordable than an outside system, and is relatively easier to service and maintain. Provide a sump pump of sufficient capacity with automatic and manual controls [3] (BS 8102:2009, BS EN 752:2008).	Ensure that access to the basement should have a crest level of min. 150 mm higher than the platform level to segregate catchments. Runoff from the roofs, rainwater downpipes and all premises at and above ground level should be channelled into surface gravity drains. Provide cut-off drains across the access way [17].	Conduct quarterly inspection to check the condition of the drainage system. Ensure the functioning of emergency maintenance services (e.g. pumps) and automatic alarm systems that notify pump failure. Clean drains and sump 4 times per year and. if possible, clean internal cavity twice a year.
Penetrations for services  Efflorescence and rust stains found just below a pipe penetration	Service penetrations are weak points and are vulnerable to leakage. They should be grouped, pre-planned and boxed out to minimise penetration through waterproofing. Install penetrations in cast-in-situ sleeves to allow independent movement of pipes and to reduce coordination between different trades (BS 8102:2009, SS CP 82:1999).	Ensure there are no congestions around pipes for easy pouring and vibration of concrete. Additional reinforcement is required to counteract concentration of shrinkage stress especially at corners of openings (BS 8102: 2009, SS CP 82:1999). Conduits should be made in order not to allow water leaks in the basement.	Routinely inspect locations of pipe penetrations, such as service entries, which are particularly vulnerable to water leakage. Keep access points clear and free from obstructions to avoid excessive hacking of finished surfaces. Check for differential movement of the pipe (BS 8102:2009, SS CP 82:1999).

FIRE PROTECTION IN BASEMENT

Mini-jet fan

Fire Hose System

Sprinkler System

Design

- Design basement slab and wall covers with fire-resistance compliance as per SS EN 1992-1-1:2008 and SS EN 1992-1-2:2008(2015).
- Specify the installation of fire sprinkler system in basement car park as per BS EN 12845:2015 and SS CP 52:2004.

Construction

- Comprehensive information and records relating to fire and security hand over as per provisions of SS EN 1992-1-1:2008.

Maintenance

- Daily cleaning and removal of furniture, equipment, stacked material, etc. that obstruct access [18].
- Monthly cleaning of fire hose (both internally and externally), kept dry, and compactly rolled.
- Monthly checking of drum hose, nozzle, stopcock and cabinet conditions for (e.g.) corrosion, leakage, etc. Provide lubrication as required.
- Monthly checking of the condition of parts (e.g. valves, glands, etc.) and functionality of each role. Any misuse/unintended use should be prevented [19].
- Monthly testing for water flow pressure.
- Yearly hydrostatic test to detect any defect or leak, especially if the hose has been exposed to chemicals or severe stress.

References

[1] International Building Code (2006). Chapter 5: General Building Heights and Areas. Retrieved on May 9 from https://codes.iccsafe.org/public/chapter/content/4188/.

[2] New York Department of City Planning (2017). Glossary of Planning Terms. Retrieved on May 9 from http://www1.nyc.gov/site/planning/zoning/glossary.page.

[3] Chew, M. Y. L. (2017). *Construction Technology for Tall Buildings* (5th ed.). Singapore: World Scientific.

[4] SPRING Singapore (1999). CP 82: Code of practice for waterproofing of reinforced concrete buildings. Singapore.

[5] SPRING, Singapore (2002). *SS 500: Elastomeric wall coatings*. Singapore: SPRING.

[6] Kubal, M. T. (2008). *Construction Waterproofing Handbook* (2nd ed.). New York: McGraw Hill.

[7] Chew, M. Y. L. (2016). *Maintainability of Facilities: Green FM for Building Professionals* (2nd ed.). Singapore: World Scientific.

[8] NUS Maintainability of Buildings (2017). Basement. Retrieved on March 9 from http:// www. hpbc.bdg.nus.edu.sg/?page_id=865&page=3.

[9] Building and Construction Authority (2003). *Waterproofing for Internal Wet Areas* (2nd ed.) Singapore: Building and Construction Authority (BCA).

[10] Public Utilities Board (2016). Flood Protection Measures, Flood Management. Retrieved on January 26 from www.pub.gov.sg/drainage/floodmanagement/floodprotectionmeasures.

[11] Public Utilities Board (2013). *Code of Practice on Surface Water Drainage* (6th Adden.). Singapore: Public Utilities Board (PUB).

[12] Public Utilities Board (2013). *Managing Urban Runoff — Drainage Handbook* (1st ed.). Singapore: Public Utilities Board (PUB).

[13] Building and Construction Authority (2004). *Good Industry Practices — Waterproofing for External Wall* (2nd ed.). Singapore: Building and Construction Authority (BCA).

[14] Building and Construction Authority (2017). *CONQUAS® The BCA Construction Quality Assessment System* (9th ed.). Singapore: Building and Construction Authority (BCA).

[15] Land Transport Authority (2009). *Materials and Workmanship Specification for Architectural Work*. Singapore: Land Transport Authority (LTA).

[16] Public Utilities Board (2004). *Code of Practice on Sewerage and Sanitary Works* (1st Adden.). Singapore: Public Utilities Board (PUB).

[17] Land Transport Authority (2010). *Civil Design Criteria*. Singapore: Land Transport Authority (LTA).

[18] Workplace Safety and Health Council (2016). *Workplace Safety and Health Guidelines on Workplace Housekeeping*. Singapore: Workplace Safety and Health Council (WSHC).

[19] Singapore Civil Defence Force (2013). *Code of Practice for Fire Precautions in Buildings*. Singapore: Singapore Civil Defence Force (SCDF).

Normative References/Standards Referred to for Basement

- ASTM C33/C33M — 16e1-Standard Specification for Concrete Aggregates
- BS 1881-210:2013 — Testing hardened concrete. Determination of the potential carbonation resistance of concrete. Accelerated carbonation method
- BS 6093:2006+A1:2013 — Design of joints and jointing in building construction. Guide
- BS 6150:2006+A1:2014 — Painting of buildings. Code of practice
- BS 6213:2000+A1:2010 — Selection of construction sealants. Guide
- BS 6398:1983 — Specification for bitumen damp-proof courses for masonry
- BS 8000-0:2014 — Workmanship on construction sites. Introduction and general principles
- BS 8004:2015 — Code of practice for foundations
- BS 8102:2009 — Code of practice for protection of below ground structures against water from the ground
- BS 8204-1:2003+A1:2009 — Screeds, bases and in-situ floorings. Concrete bases and cementitious levelling screeds to receive floorings. Code of practice
- BS 8204-2:2003+A2:2011 — Screeds, bases and in-situ floorings. Concrete wearing surfaces. Code of practice
- BS 8210:2012 — Guide to facilities maintenance management
- BS 8215:1991 — Code of practice for design and installation of damp-proof courses in masonry construction
- BS 8221-1:2012 — Code of practice for cleaning and surface repair of buildings. Cleaning of natural stone, brick, terracotta and concrete
- BS EN 12845:2015 — Fixed firefighting systems. Automatic sprinkler systems. Design, installation and maintenance
- BS EN 13139:2002 — Aggregates for mortar
- BS EN 14630:2006 — Products and systems for the protection and repair of concrete structures. Test methods. Determination of carbonation depth in hardened concrete by the phenolphthalein method
- BS EN 1504-2:2004 — Products and systems for the protection and repair of concrete structures. Definitions, requirements, quality control and evaluation of conformity. Surface protection systems for concrete
- BS EN 1504-3:2005 — Products and systems for the protection and repair of concrete structures. Definitions, requirements, quality control and evaluation of conformity. Structural and non-structural repair
- BS EN 1504-5:2013 — Products and systems for the protection and repair of concrete structures. Definitions, requirements, quality control and evaluation of conformity. Concrete injection
- BS EN 1504-9:2008 — Products and systems for the protection and repair of concrete structures. Definitions, requirements, quality control and evaluation of conformity. General principles for use of products and systems

- BS EN 15651-3:2017 — Sealants for non-structural use in joints in buildings and pedestrian walkways. Sealants for sanitary joints
- BS EN 1992-1-1:2004+A1:2014 — Eurocode 2: Design of concrete structures. General rules and rules for buildings
- BS EN 1992-1-2:2004 — Eurocode 2. Design of concrete structures. General rules. Structural fire design
- BS EN 1992-3:2006 — Eurocode 2. Design of concrete structures. Liquid retaining and containing structures
- BS EN 752:2008 — Drain and sewer systems outside buildings
- BS EN ISO 11600:2003+A1:2011 — Building construction. Jointing products. Classification and requirements for sealants
- SS 150:2015 — Specification for emulsion paint for decorative purposes
- SS 509-1:2015 — Code of practice for cleaning and surface repair of buildings — Part 1: Cleaning of natural stone, brick, terracotta, concrete and rendered finishes
- SS 542:2008 — Code of practice for painting of buildings
- SS CP 4:2003(2012) — Code of practice for foundations
- SS CP 52:2004 — Code of practice for automatic fire sprinkler system
- SS CP 65-1:1999 — Code of practice for structural use of concrete — Design and construction
- SS CP 82:1999 — Code of practice for waterproofing of reinforced concrete buildings
- SS EN 12620: 2008 — Specification for aggregates for concrete
- SS EN 1992-1-1:2008 — Eurocode 2: Design of concrete structures, Part 1–1 General rules and rules for buildings
- SS EN 1992-1-2:2008(2015) — Eurocode 2: Design of concrete structures, Part 1–2 General rules — Structural fire design

Chapter

Wet Areas

Introduction

Wet areas refer to areas which are constantly subjected to the presence of moisture. In this context, internal wet areas define areas like bathrooms, toilets and laundries.

Special attention must be paid to wet area flooring due to its exposure to frequent interchanging dry and wet cycles, which may cause unhealthy or hazardous conditions [1]. Most of the problems in wet areas stem from the failure to achieve internal water-tightness. Waterproofing membranes are applied on the floors and walls in order to make wet areas watertight. Although the waterproofing of wet areas is not subjected to hydrostatic pressure, it contains numerous pipe penetrations, projections and structural joints to meet the areas' functional and aesthetic requirements. Continuity of waterproofing suffers at pipe penetrations and structural joints between the floor and a wall. The walls adjacent to fixtures such as the shower also require waterproofing because these fixtures are the source of splashing and seepage of adjacent walls and floors.

The annual maintenance cost for wet areas can range from 35% to 50% of the total maintenance cost of a building [1], thus it is required to consider the maintainability aspects of wet areas at the design stage. The identified defects of wet areas, and the corresponding design, construction and maintenance guidelines to tackle each defect, are presented in the succeeding tables. The structural defects relate

Artist's impression of a bathroom in a private residential unit.

to seepage through structural joints, leakage through concrete slabs, leakage through walls/floors, and corrosion and concrete spalling issues. The architectural defects pertain to efflorescence, biological stains, and issues concerning paint, tile interface with other elements, tile cracks and staining, tile joints, floor gradient and screed, and waterproofing. The service defects are concerned with wet area access, penetrations for sanitary fittings and issues regarding fixtures and fittings.

Aside from durability and aesthetic considerations, the success of design decisions pertaining to wet areas depends on the compatibility of all systems and components. Designing and installing a well-organised plumbing layout will reduce the requirement for penetrations through waterproofing membrane, and will provide easy access for inspection as well as easy cleaning. The design and installation of an adequate slope in the floor of a wet area is required for proper drainage. Screed design also requires careful consideration of fitting the wet area's layout, to accommodate the recommended falls and services for plumbing. Sanitary fixtures and fittings should ensure a controlled water flow and must not contribute to defects (e.g. leakage, seepage, splashing, difficult access, etc.) in order to keep the wet area hygienically clean and dry. The design of window openings or mechanised ventilation should provide proper ventilation to prevent biological growth on floors, ceilings and walls due to inadequate air circulation [1].

The substrate structure's design and area of coverage should be considered in selecting a specific waterproofing system to ensure the water-tightness of the wet area unit. A well-prepared surface, quality material and good workmanship are the basic requirements for waterproofing, as well as the thorough protection of installed waterproofing membrane can augment its useful life to a significant extent [2]. Compatibility with the substrate type and thickness of the tile and bedding is required in selecting the correct type of bedding material for tiling in order to minimise defects (e.g. debonding, cracking and efflorescence). Movement joints, tile sizes, grouting materials and pointing should be considered at the design stage to minimise defect occurrences during construction and post-occupancy. Movement joints accommodate stresses due to shrinkage, deflection and moisture, while good quality grouting materials seal up tile joints and prevent water seepage. Paint selection considerations during the design stage is crucial since the paint film will be constantly subjected to moist conditions leading to peeling, flaking, blistering, biological attacks and efflorescence [3].

Wet areas deteriorate with time and need maintenance. Accessibility for repair and the replacement of service pipes should be dealt with during the design stage. Provision of easily opened covers and walk-in pipe ducts will help expedite maintenance access. Positions of service pipes should be considered in the design stage so that the pipes do not inhibit cleaning accessibility of the entire floor and wall areas. Regular cleaning using correct cleaning methods are important considerations in wet area maintenance programmes. Unforeseen stains due to countless human and goods movements and or extended surface contact with particular solutions may require special cleaning.

2.1 Structural

The structural components of a building (e.g. foundation, floor, slab, walls, and other load bearing and non-load bearing components) require design, construction and maintenance specifications to prevent premature deterioration and help maximise the wet area's performance.

Problem	Design	Construction	Maintenance
Seepage through structural joints  Efflorescence due to seepage along joints  Water stains and flaking	Minimise the provision of structural or construction joints to maintain monolithic construction (BS EN 1992-3:2006). Provide special reinforcement detailing at the joints and 1500mm lapping. Fibreglass mesh at joint is also recommended, (BS 6093:2006+A1: 2013, BS EN 1992-1-1:2004+A1:2014). Specify compatible membrane that effectively accommodates for movement (SS CP 82:1999).	Check substrate readiness prior to installation and dampen surface to receive cementitious membrane (SS CP 82:1999). Mix the correct proportions of waterproofing product until it is homogeneous and lump free. Lay a protective slurry coat above membrane and cover any openings to protect from debris. Lay protective screed with sufficient fall (BS 8204-2:2003+A2:2011, BS 8000-0:2014, BS 8000-9:2003) [4].	Conduct regular inspections of wet area, especially at vulnerable interfaces/joints between different materials. Use non-destructive testing (NDT) in this regard. Water leakage can be identified using thermography images (BS ISO 10880:2017) or by using a moisture meter (BS 812-109:1990) to determine the presence of moisture. Repair work can be done using polyurethane (PU) grouting/injection for local seepages. For more severe incidents, laying of new waterproof screed is strongly recommended [5].
Leakage through concrete slab  Leakage on porous slab	Compliance on structural design as per SS EN 1992-1-1:2008 to address strength, integrity, porosity, etc. For water-tightness conditions, comply as per SS CP 82:1999. Include movement joint with proper detailing as per BS EN 1992-1-1: 2004+A1:2014, SS EN 1992-1-1:2008.	Maintain sufficient cover for reinforcement bars with proper compaction and curing (SS EN 1992-1-1:2008, BS EN 1992-1-1:2004+A1: 2014, BS EN 1992-3:2006).	Avoid water infiltration into the structure by keeping the concrete slab as dry as possible, using good housekeeping practices. Provide protective measures to prevent waterproofing membrane from damage by any activities on site (SS CP 82:1999).

(*Continued*)

(*Continued*)

Problem	Design	Construction	Maintenance
Leakage through cracks and porous walls/floors  Efflorescence and corrosion  Blistering and biological growth through RC internal floor slab due to leakage	Design specifications for appropriate cement and reinforcement for concrete work in wet areas. Proper design of waterproofing to prevent seepage through wall and slab, especially at vulnerable points such as penetrations [4]. Detailing to avoid embedded pipes in floors/walls to maintain sufficient cover to reinforcements [5]. (See also BS 8215:1991). Ensure the segregation of dry and wet zones [1]. Specify proper mix design and adequate mixing of concrete[1] (BS EN 1992-1-1:2004+A1:2014, SS EN 1992-1-1:2008).	Maintain water–cement ratio so as to avoid excessive shrinkage. Adhere to recommended curing process (BS EN 1992-1-1:2004+A1:2014, SS EN 1992-1-1: 2008). Ensure sufficient reinforcement cover and proper vibration of concrete to avoid honeycombs [6]. Ensure accurate batching of fresh concrete and full compaction at the earliest possible time. Conduct water ponding test; flood waterproofed area (25mm depth) for 24 hours, inspect deck below for leakage) [5].	Perform regular inspection for signs of water seepage using NDT diagnostic techniques. Conduct monthly cleaning and timely repair (SS CP 82:1999). Remediate any leakage by removal of screed, clearing all loose particles and re-application of waterproofing. Localised porous concrete can be repaired by PU grouting by injecting into either the passive or active side of theslab/wall. Repair leakage at cracks by either injecting polyurethane (PU) grout, or by using the *Flood Infusion Method* [7].
Corrosion and spalling of concrete  Corrosion of rebars on slab 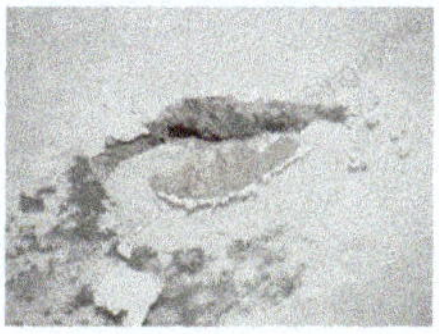 Spalling and flaking	Adhere to the proper concrete design and specification of rebar (SS EN 1992-1-1:2008, BS EN 1992-1-1:2004+A1:2014, BS EN 1992-3:2006). Specify use of admixtures as per SS EN 934-2:2015. Design of concrete must prevent cracking/spalling of walls at points of support due to differential settlement. Design of concrete must inhibit cracking, spalling, and local bulging of non-load bearing and partition walls due to thermal and moisture movements of partitions or supporting structures (ISO/NP 4356).	Maintain sufficient cover for reinforcement bars with proper compaction and curing (SS EN 1992-1-1: 2008, BS EN 1992-1-1:2004 +A1:2014, BS EN 1992-3:2006). Ensure sufficient ventilation to reduce the humid environment that speeds up carbonation [8]. Assess cracking (as a precursor to spalling) in the compression region of concrete structural components, since such cracks form parallel to the principal compression stresses which may signal an increased damage severity (ISO 28841:2013).	<u>Preventive:</u> Paint concrete every 3 to 5 years using anti-carbonation or good quality paint to prevent carbonation. Check regularly for any holes or cracks and seal up any holes immediately to prevent moisture and carbon dioxide from entering the concrete [8]. <u>Corrective:</u> All loose and flaking material should be cut back to a firm base (SS 509-2:2005(2015), BS 8221-2:2000). Corrosion and spalling must be attended to immediately [8] by means of patch repair with approved polymer modified sand cement mortar.

(*Continued*)

[1] Reduce biological growth by lowering occurrence of shrinkage cracks in concrete by using cement and aggregate with low shrinkage characteristics (as per SS EN 12620:2008), using admixture to control shrinkage (as per SS EN 934 Series — Admixtures for concrete, mortar and grout), using low water content, and maximising coarse aggregate content.

OTHER EXAMPLES OF WET AREA DEFECTS [2]

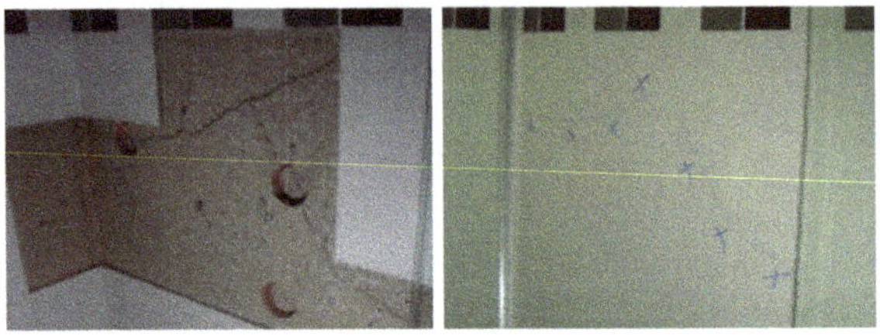

Examples of Crazing

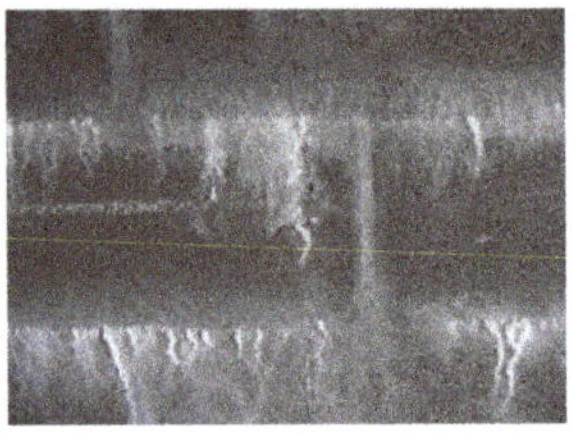

Example of Crystallization damage

Crazing — results from differences of moisture movement due to high moisture concentration gradient or a discontinuity in compaction near the exposed surfaces. The cracks rarely exceed 12mm or so in depth, and are not serious, apart from their unsightliness.

Crystallisation damage — Sulphate or other salts may be admitted by permeable concretes and if, at a later stage, drying occurs, these salts crystallise. The condition may be severe when concrete is subjected to wetting and drying cycles, or if one area of the structure is saturated while the adjacent area is dry. Salts migrate toward the dry area, causing extensive crystallisation in the region of drying.

2.2 Architectural

Architectural components define the character of the building. They include the finishes, furnishings, and other elements that contribute to the building's aesthetic value and liveability. The design, construction and maintenance specifications for the wet area's architectural components will ensure the optimal performance and better quality workmanship of the building.

Problem	Design	Construction	Maintenance
Efflorescence  Efflorescence on floor tiles  Efflorescence on wall tiles	Enforce quality control for cement and curing (BS EN 1992-1-1: 2004+A1:2014, SS EN 1992-1-1:2008). Refer to BS EN 771-1: 2011+A1:2015 for the use and limits of soluble salts for clay bricks. Specify grout admixtures that inhibit efflorescence. Specify grout/mortars complying with ISO 13007-1, ANSI A118.15.	Store construction material on site, off ground, with some form of covering as to protect them from moisture ingress. Avoid wetting if possible as this will eventually trigger future efflorescence outbreak [9]. Clean and free all intended mortar joints of debris, and fill them completely; avoid cavities and air spaces. Use mechanical vibration to	Conduct regular cleaning of efflorescence as per SS 509-1:2015 (BS 8221-1:2012). New efflorescence stains can be removed with water and a scrub brush, while older stains may require water blasting or sandblasting. For difficult stains, the use of hydrochloric or phosphoric acid is recommended. However, as chemical solutions can discolour the

(*Continued*)

(*Continued*)

Problem	Design	Construction	Maintenance
 Efflorescence on marble finish	Prevent water penetration of the exterior wall by providing waterproofing and architectural details (e.g. isolation of exterior brick wythe with an air cavity. If no cavity wall, separate the brick wythe from the backing with a damp-proof coating) (ASTM C1400-11).	consolidate the grout to reduce voids. Use dense tooled mortar joints to impede salt migration. Install sealant joints between masonry and door/window frames, expansion joints and other vulnerable interfaces (ASTM C1400-11).	affected area, all visible concrete in the area is cleaned to maintain a consistent appearance[2] [10]. Rinse surface thoroughly after removal of efflorescence and carry out timely repair/maintenance of leakage to avoid future efflorescence.
Biological stains  Algae growth at seepage  Traces of fungi and mould stains 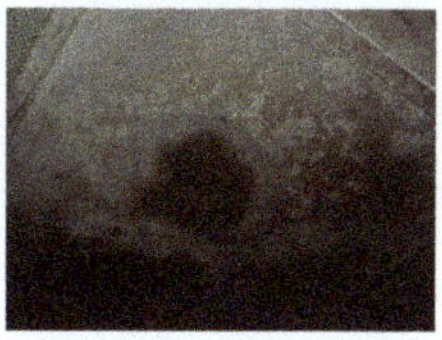 Biological stains on ceiling	Detailing of wall surfaces in terms of regularity and surface texture to prevent staining [11]. Wet areas with high water splash must be impervious with waterproofing membrane up to a minimum of 300mm (height). Also such surfaces must be without impediments to ensure ease of cleaning [7]. Provide access for adequate cleaning (BS 8221-1:2012, SS 509-1:2015). Selection of suitable algae resistant paint products conforming to the checklists for paint work published as per Building and Construction Authority (2001), *Good Industry Practices – Painting*.	Exterior surfaces of porous building material (e.g. brick, stone, cement rendering) can develop biological growth (e.g. mosses, lichens, algae). Any such growth should be avoided as much as possible with treatments of anti-algae/anti-fungus solutions and allowed to dry before painting/ repainting. Improve ventilation and remove sources of dampness to dry out the walls as thoroughly as possible during construction/painting works (BS 6150:2006+ A1:2014, SS 542:2008).	Conduct regular cleaning by chemical and mechanical applications (e.g. alkaline or solvent cleaning agents, hydrofluoric acid, air or water abrasive cleaning) and clear dirt regularly as per SS 509-1:2015. (See also BS 8221-1:2012). Remove moulds, lichens and other growths with a stiff brush and treat the residue with biocide chemicals [12]. Use paint which does not support mould growth. If affected, remove infected paint and sterilise the surface by applying an antiseptic wash to prevent recurrence (BS 6150:2006+ A1:2014, BS EN ISO 1513:2010, SS 542:2008).

(*Continued*)

[2] See *Maintainability of Buildings, Efflorescence. Available at: http://www.hpbc.bdg.nus.edu.sg/?page_id=8228 (accessed May 12, 2017)* for more on source, composition and removal of efflorescence.

(*Continued*)

Problem	Design	Construction	Maintenance
Paint  Peeling and flaking 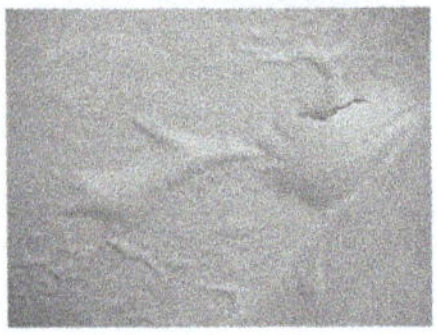 Blistering	When selecting paint, consider substrate, environment, application method, feasibility of surface preparation, over coating interval, and appearance (BS 6150:2006+A1: 2014, SS 542:2008). Ensure that selected paint complies with the water resistance requirements (Water based emulsion paint; cement based paint system; water repellent coatings) (BS 6150: 2006+ A1:2014, SS 542:2008).	For painting (protective and decorative), apply the relevant sections of BS EN ISO 12944-4:1998 and BS 6150:2006+A1:2014 and ensure that substrates are sound before commencing painting work. Completed paint work should be cordoned off and protected until adequately dried to avoid stains and other damages. Adhesion and visual properties to be consistent with the approved sample during final inspection [13].	Complete all repairs of damages due to wear and tear by washing down and removing defective paint film. Apply sealer/primer (if necessary), and repaint (BS 6150:2006+ A1:2014, SS 542:2008). Conduct regular cleaning as per site condition. Avoid excessive moisture exposure on painted surface to avoid mildew or fungus formation. Wash minor stains and dirt with clean water, and wash heavy stains with mild detergents.
Tile interface with other elements  Inadequate sealant creates water pockets between and under tiles	Specify an adhesive that is cured by hydration and avoid the use of dispersion adhesion over impervious tiles as it is highly unlikely that the adhesive will achieve full curing (BS EN 12004-1:2017). In wet areas, the minimum adhesive coverage of tiles should be at least 90% (BS 5385-1:2009, SS CP 68:1997).	Maintain consistent tile interface with ceiling, window frames, door frames, pipes, etc. [14] (BS 8000-0:2014) Adhesive coverage is critical at edges of tiles. Full coverage would eliminate the formation of water pockets under the tiles (BS EN 12004-1:2017, BS 8000-9:2003, BS 8000-11:2011).	Visually inspect for consistency (SS CP 82: 1999). Conduct random checking of bedding by removing tiles and assess fault distribution to avoid voids left behind tiles (BS 5385-1:2009, SS CP 68:1997).

(*Continued*)

(*Continued*)

Problem	Design	Construction	Maintenance
Tile cracks  Crack on floor tile  Crack due to uneven or hollow surface 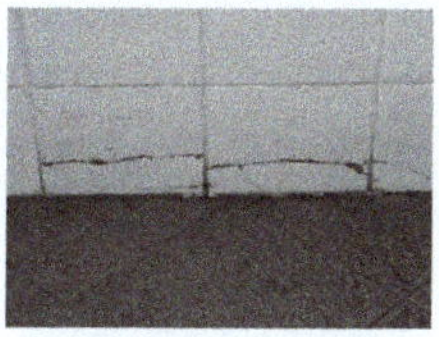 Crack lines due to traffic damage and lack of skirting along the wall–floor tile interface	Design to accommodate differential movements of the structure and provide movement joints to mitigate differential settlement or shrinkage of cracks. Recommended minimum wall joint width of 3mm (5mm preferred) and minimum floor joint width of 5mm. Spacing of movement joints on internal wall should be 5–6m horizontally and vertically, while space for internal floor should be 6–7m in all directions (BS 5385-1:2009, SS CP 68:1997). Specification of less water-permeable tiles for wet areas [15]. Recommend testing of tiles against design specifications before mass installation (ISO 10545-1:2014). Cracks at tile joints are commonly due to the settling process, and thus requires pointing (BS 5385-3:2014).	Comply with the tile fixing method as per SS CP 68:1997 (BS 5385-1:2009, BS 8000-11:2011), including: • Careful cutting and handling of tiles during application • Soak tiles to relieve its dry condition before laying • Screed needs to be properly cured, rendered and cleaned • Tiles should not be laid over cracks or be subjected to any loading • Tiles should be properly tapped in place, and have their surfaces wiped after tiling Tiles should resist cracking from soaking or transverse elongation. They should be of a thickness adequate for resisting cracks from direct impact. Use a proper key at the back of the tiles for good bonding.	Conduct surface cleaning and repair applications. Use a stiff brush to loosen joint grout and reinstate joints with suitable tile grout. Remove and re-tile the affected areas (SS CP 82:1999); (there may be a colour mismatch if stocks of extra tiles are unavailable). For areas where the wear is negligible, a mixture of cement and water slurry with a hardener may be painted over them. However, for large worn surfaces, it is necessary to hack off the screed and patch up with a suitable mix of cement and sand screed together with a hardener (BS 8221-2:2000). Perform regular mopping/buffing of tiled surfaces. Proper handling of equipment to prevent damages to the tile surface.

(*Continued*)

(*Continued*)

Problem	Design	Construction	Maintenance
Tile staining  Efflorescence caused by rising dampness, leading to migration of soluble salts in the substrate/grout 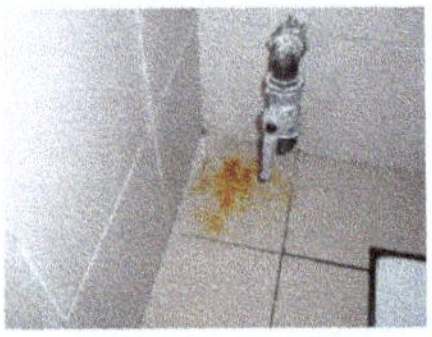 Corrosion on floor tiles due to water dripping from faucet  Rust stain on wall tiles 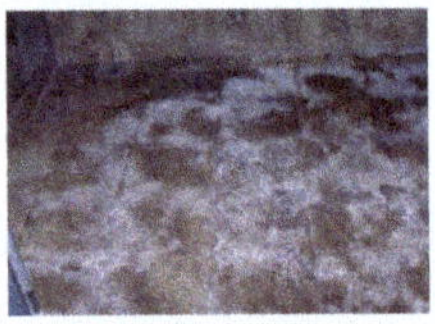 Staining of floor tiles	When selecting tiles, consider the following criteria: impermeability (resistance against water absorption); slip resistance (both dry and wet); impact and abrasion resistance; and resistance against chemicals, dirt and stains (ISO/DIS 13006). The use of pre-packed mortar is preferred with proper mixing [16]. Select and specify the use of stain resistant tiles (ISO 10545-1:2014). Specify for allowable levels of unevenness of tiled surfaces as per SS CP 68:1997 (see also BS 5385-1:2009). Specify for larger tiles to reduce the incidence of grouting. Design for correct grouting width between tiles. Alternatively, use polymer-modified grout to reduce risk of shrinkage cracks.	Pre-packed mortar mix is recommended for consistency. Adhere to recommended mixing proportions. Use mechanical mixers prior to screeding. Rigorously supervise on-site work for high quality workmanship (BS 8000-0:2014). Use suitable adhesive according to supplier's recommendations [16]. Check tile installation for consistency in size, thickness of skirting, pointing, alignment of joints, evenness, cleanliness and colour tone of surface, and any signs of chipping, cracking, or hollowness. Once the tiles have been installed, immediately cover them with polythene sheets, cardboards or wooden boards to protect the surface from foot traffic, abrasion, and impact [17].	Mopping and buffing of commercial areas should be thrice a day while other areas should be twice a day. Fortnightly, machine scrub floors to remove grout stain. Perform monthly hand scrubbing of walls. Use commercial tile cleaner to remove efflorescence twice a year. Use diluted chlorine bleach/mildew-retardant spray for mildew, and in areas of deep soiling and hard to reach areas. However, the source of staining should first be removed (BS 8221-1:2012, SS 509-1:2015). Biological growth can be removed by scrubbing with an acid/alkali based cleaner. Biocidal washes may be applied to prevent further growth. Remove severe soiling by scrubbing with acid/alkali based cleaner.[3]

(*Continued*)

[3]See *Maintainability of Buildings, Ceramic Tile Cleaning. Available at: http://www.hpbc.bdg.nus.edu.sg/?page_id=8831 (accessed May 12, 2017)* for more information about cleaning special tile stains caused by various human activities and prolonged exposure of certain solutions.

(*Continued*)

Problem	Design	Construction	Maintenance
Tile joints 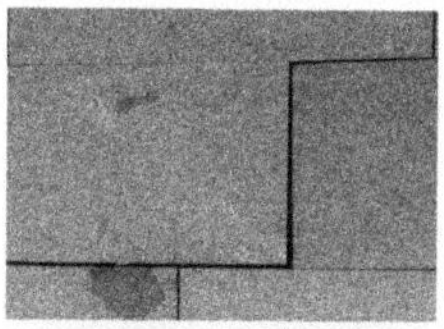  Uneven tile jointing 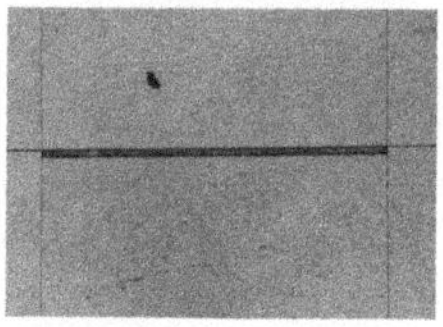 Tile gap between floor tiles	Ensure that the joint location, tile's width and substrate are consistent. The maximum recommended spacing for internal wall and floor is 5–6m; while the minimum joint width is 3–5mm for wall and 5mm for floor. Refer to movement joint guidelines as provided by SS CP 68:1997 (BS 5385-1:2009). Specify grout which are compatible with the tiles and which possess resilience and compressibility [17]. Provide detailed specifications of surface regularity and tolerance levels for quality control.	Grouts should be of suitable fineness and consistency according to the joint width stipulated in BS 5385-3: 2014. Samples of grouts need to be tested at an accredited laboratory.[4] Joints should be aligned and of a consistent size (between 1–2mm) [16]. Ensure joints are pointed neatly with no voids within them and no excess or uneven grout (BS 5385-1:2009, SS CP 68:1997). Tiling should be divided into bays at all expansion/ structural joints and points of stress concentration [14].	Perform surface cleaning so tiles are dry, free from dust and foreign materials (SS CP 82:1999). Remove affected tiles by saw cutting grouted joints. Clean adhesive off the tile surface before it sets in, without disturbing the tiles. Use a stiff brush to loosen joint grout and reinstate joints with suitable fresh tile grout. Ensure substrate surface is free from loose particles and of correct level before replacement of new tiles (BS 5385-1:2009, SS CP 68:1997).
Floor gradient and screed  Water puddling in bathroom due to uneven surface gradient	Screed provides protection to the waterproofing membrane and constructs the slope for drainage [11]. Provide adequate thickness to embed the service pipes by at least 20mm. For thicker screed, coarse	Ensure quality control of workmanship as per BS 8000-0:2014 for proper pouring, curing and compaction. Concrete floors must be air-dried for a minimum of 4 weeks after curing. Slope should be maintained using a series of spot levels. Special attention	Conduct water-tightness test to check screed performance in terms of dampness, curing and thickness of screed (SS CP 82:1999). Check integrity of screed and perimeter divorcement.

(*Continued*)

[4] See *Maintainability of Buildings, Grouting to Tile Joints. Available at: http://www.hpbc.bdg.nus.edu.sg/?page_id=8851 (accessed May 12, 2017)* for the characteristics of grouts, test methods (ANSI) and performance criteria.

(*Continued*)

Problem	Design	Construction	Maintenance
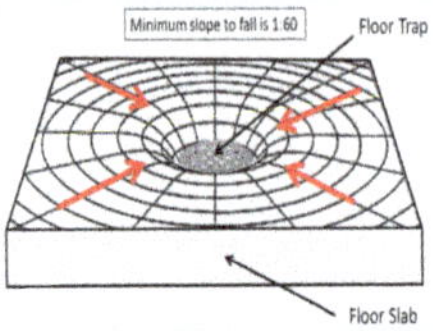 Minimum slope to fall details	aggregates of a smaller size should be used. Increase drop considerably in cases where the layout requires soil pipe to be embedded in the screed, to maintain a minimum screed depth [18].	must be given to the installation of all penetrations and movement joints. The pipes embedded in the screed should be checked for proper layout and jointing [5]. Movement joints in structural slab should be carried through the screed (BS 8204-1:2003+A1:2009).	Ensure there is no obstruction to flow of water, so that water will drain off along the gradient (SS CP 82:1999). Regularly inspect for initial signs of water seepage. If the space is above ground floor, check the space or ceiling directly under for any sign of water damage [19].
Waterproofing  Efflorescence due to water ponding at the corner 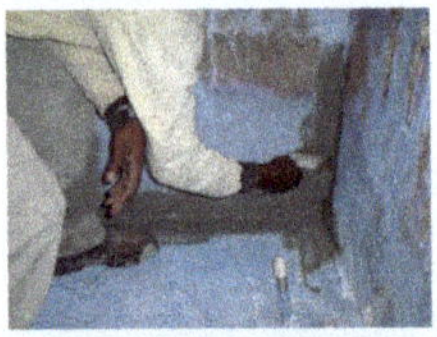  Proper waterproofing application to ensure that wall and floor joints are well-sealed	Select waterproofing material and check for required material properties as per related standards (SS CP 82:1999). For wet walls in showers/bathrooms, waterproofing should be applied to at least 1500mm (width) and 1800mm (height) of the wall. If a basin or sink is within 75mm of the wall, the wall adjacent to and behind it should receive waterproofing for a minimum height of 300 mm. Specify fibreglass/ reinforcements/ fillets at wall-floor interfaces [5]. Specify product requirements for liquid-applied waterproofing membrane products (ISO 13007-5:2015).	Surface prior to waterproofing membrane application should be even, with fine roughness and without sharp corners or protrusions. Surface should be thoroughly cleaned and dried such that moisture content is <6%. [13]. When applying the waterproofing, begin from the corner furthest away from the entrance and work towards the door so as to avoid stepping on the freshly applied membrane. Ensure minimum upstands and lapping; and the laying of fibreglass/ reinforcements/fillets at corners as per design specifications. Add a slurry coat to protect the membrane against any damage at installation [5].	Conduct a full inspection (for seepage, dampness, relative humidity (RH), mould growth, soiling, corrosion, etc.) of wet area at least once a year. Suggest using thermographic imaging to detect any sub-surface leakage. Conform to the maintenance practices guidelines for cleaning of finishes, fixture and fittings (BS 8221-1:2012, SS 509-1:2015, BS 6270-3:1991), Check adhesion using the pull off test. Average tensile pull out strength of the five spots should be ≥ 0.40 N/mm2; and the individual pull-out strength of each sample should be ≥ 0.30 N/mm2 [5]. Conduct water-tightness test for floor (ponding test) [14].

2.3 Services

Service components refer to the vertical and horizontal circulation systems, electro-mechanical and sanitary connections in buildings. Each issue is presented with its corresponding design, construction and maintenance measures to ensure better performance. Note: The design and installation of services should follow the relevant sections in the Code of Practice on Sewerage and Sanitary Works, and SS CP 48; Code of Practice for Water Services. See also BS 8542:2011, BS 8554:2015, BS 8595:2013 and BS 8000-0:2014.

Problem	Design	Construction	Maintenance
Wet area access  Access panel inadequately located; fails to accommodate maintenance of concealed plumbing  Exposed as opposed to concealed plumbing	Locate stack risers in easily accessible areas for maintenance. Consider exposed services when aesthetically possible. Access panels should be provided at adequate locations and in the right size, such that all concealed plumbing are fully serviceable. Specify the location of utensils to be highly accessible for ease of cleaning, and which impose minimal obstructions for cleaning and maintenance of the walls and floor [11].	Install access panels that are large and central enough to accommodate the movement required for maintenance; in most inconspicuous manner. Ensure all pipework are concealed, except final connections to fixtures [20]. Proper installation of pipe joints in concealed areas to provide prolonged water-tightness without need for maintenance. After installation and curing, conduct a full pressure test on pipe work to ensure no leaks from joints before embedding into concrete and use.	Adopt an efficient inspection system for the wet area[1] and conduct regular inspection of the plumbing system to identify any defects (BS 8210:2012). Any sign of water seepage should be further investigated and quickly resolved before associated damage can occur to the wet area. Ensure serviceability of concealed services by routinely inspecting their access panels for accessibility.

(Continued)

(*Continued*)

Problem	Design	Construction	Maintenance
Penetrations for sanitary fittings (piping layout)  Leak at pipe joint  Loose tile cladding due to service pipe intrusion from slab through wall and beam	Provide a well-planned layout and detailed working drawing to reduce chances of hacking or porous infill of cold joints (BS EN 12056-2: 2000). Number of penetrations should be minimised by using common discharge stacks and cast-sleeve[5]: • For WC/Urinal — 1 trap for a max. of 10 urinals. • For WB — 1 washbasin trap to serve 10 WBs. More than 1 trap provided if over 10 WBs. • For Wash/shower/ bath — 1 floor trap for every 3 WC cubicles. Two more penetrations for common stacks (drainage and vent).	The membrane should be dressed up at pipe penetrations and down at least 50mm into the floor outlet. Use of reinforcing fibreglass mesh along with the upstand is a better solution. The membrane should extend horizontally around the pipe by min. 100mm and overlap with subsequent membrane applied to the entire floor (SS CP 82:1999) [5]. A PVC flange should be bonded to the flooring and the PVC waste pipes before other fixtures (e.g. grates) are fitted (AS 3740:2010).	Wash and clean bathroom fittings regularly with mild detergents and remove any solid waste that may cause choking. Check pipe fittings regularly and, if necessary, have them repaired by a PUB licensed plumber [21]. Check for water-tightness. Give careful consideration to possible openings and penetrations in fittings [22]. Remedy penetrations on pipe entries using local grouting via injection packers (BS EN 1504-5:2013).
Fixture and fittings 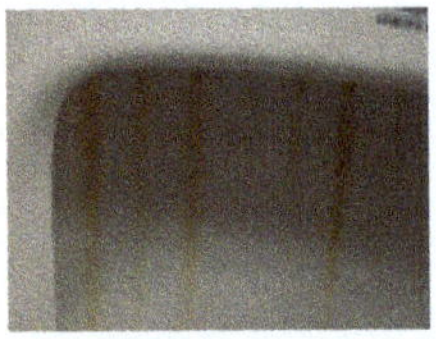 Stains on urinal  Stains on wash basin	Shower or wash area is a wet zone and should, ideally, be separated from the common toilet area (dry zone) so as to minimise the impact of water from affecting the entire floor. It should be separated by a sunken floor or kerb (min. 75 mm) [5].	Installation should comply with manufacturer's instructions or as specified by codes. No valve or faucet should leak upon installation (SS CP 82:1999). Conduct full pressure test on the pipe work upon completion to check for leaks from joints and connections. Connect and test waste water discharge pipe as per specifications [23].	Perform scheduled maintenance to check for clogged outlets (SS CP 82:1999). Recommended frequency of general (thorough) cleaning of wet areas: • Offices: 4–5 times/day • Hotels: 6 times/day • Retail spaces: 6–8 times/day Spot cleaning should be carried out the rest of the time.

(*Continued*)

[5]See *Maintainability of Buildings, Comprehensive Inspection System. Available at: http://www.hpbc.bdg.nus.edu.sg/?page_id=8499 (accessed May 12, 2017)* for recommendations on how a maintenance inspection regime can be implemented.

(*Continued*)

Problem	Design	Construction	Maintenance
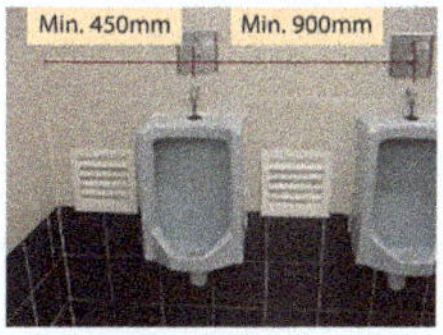 Urinal centreline distance 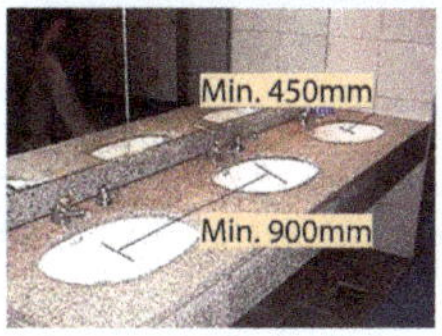 Wash basin centreline distance	For wash basins, a minimum centreline distance of 900mm, and a minimum distance of 450mm between the centreline of adjacent wall is recommended. For water closets (WC), a minimum of 450mm between the centreline of the fixture and adjacent wall or modesty board is recommended. For urinals without a partition or modesty board, a minimum centreline distance of 900 mm between adjacent fixtures is recommended [20]. Specify use of plastic pipes to minimise corrosion and subsequent joint leakages in drainage pipe system. Provide protection against corrosion to all ferrous metal components. Provide additional protection with paint coatings if required as per BS EN ISO 12944 series.	Handle fixtures and fittings carefully to avoid damaging waterproofing membrane [5]. Vanity tops/counters need to be constructed with proper slope/fall to prevent water from ponding. Materials used therein need to be selected depending on the environment and usage requirements. Vanity tops made of compressed marble are preferred as they possess relatively easy workability and the rugged durability of stone. Pipes should be fixed with relevant allowances for movement, to reduce strains during moving water that may damage joints. Further provide sliding supports for locations or rigid fixing for clamping (ISO/TS 7024:2005).	A planned maintenance programme will maintain the appearance of vanity top finishes, and fast removal of any/all staining will reduce the seriousness of the defects. To remove stubborn stains for WBs, use disinfectant cleaners and scrub with scrubbing pads weekly; while for WCs stains, use disinfectant or mild abrasive cleaners and scrub with scrubbing pads weekly. For mirrors, use neutral or ammonia-based cleaners and wipe daily. Rinse off any excess cleaners. For stainless steel/chrome fixtures, brush off any scales or rust and polish monthly. For plastic/PVC items, use neutral based cleaners for monthly cleaning. Do not scrub the surface.[6]

[6]See *Maintainability of Buildings, Wet Area. Available at: http://www.hpbc.bdg.nus.edu.sg/?page_id=84 (accessed May 12, 2017)* for more information on good practice guidelines on wet area design, construction and maintenance.

References

[1] Chew, M. Y. L. (2016). *Maintainability of Facilities: Green FM for Building Professionals* (2nd ed.). Singapore: World Scientific.

[2] NUS Maintainability of Buildings (2017). Wet Area. Retrieved on March 9 from http://www. hpbc.bdg.nus.edu.sg/?page_id=606&page=3.

[3] Carraher, C. E. Jr. (2013). *Introduction to Polymer Chemistry* (3rd ed.). New York: CRC Press.

[4] Building and Construction Authority (2004). *Good Industry Practices — Waterproofing for External Wall* (2nd ed.). Singapore: Building and Construction Authority (BCA).

[5] Building and Construction Authority (2003). *Waterproofing for Internal Wet Areas* (2nd ed.). Singapore: Building and Construction Authority (BCA).

[6] Building and Construction Authority (2006). *Precast Concrete Elements*. Singapore: Building and Construction Authority (BCA).

[7] Housing and Development Board (2017). Ceiling Leaks, Home Care Guide. Retrieved on February 22 from www.hdb.gov.sg/cs/infoweb/residential/living-in-an-hdb-flat/home-maintenance/ceiling-leaks.

[8] Housing and Development Board (2017). *Spalling Concrete. Home Care Guide.* Singapore: Housing and Development Board (HDB).

[9] Dunster, A. and Quillin, K. (2013). *Applications, Performance Characteristics and Environmental Benefits of Alkali-Activated Binder Concretes*. London: BRE Press.

[10] Woodson, R. D. (2009). *Concrete Structures: Protection, Repair and Rehabilitation.* Oxford: Butterworth-Heinemann.

[11] Building and Construction Authority (2016). *Design for Maintainability Checklist.* Singapore: Building and Construction Authority (BCA).

[12] Building Research Establishment (1992). *Control of Lichens, Moulds and Similar Growths*. London: BRE Press.

[13] Building and Construction Authority (2004). *Good Industry Practices — Painting* (2nd ed.). Singapore: Building and Construction Authority (BCA).

[14] Building and Construction Authority (2017). *CONQUAS® The BCA Construction Quality Assessment System* (9th ed.). Singapore: Building and Construction Authority (BCA).

[15] Building and Construction Authority (2009). *Design and Material Selection for Quality (Volume 2)*. Singapore: Building and Construction Authority (BCA).

[16] Building and Construction Authority (2003). *Marble & Granite Finishes* (2nd ed.). Singapore: Building and Construction Authority (BCA).

[17] Building and Construction Authority (2003). *Ceramic Tiling* (2nd ed.). Singapore: Building and Construction Authority (BCA).

[18] Building and Construction Authority (2008). D*esign and Material Selection for Quality (Volume 1)*. Singapore: Building and Construction Authority (BCA).

[19] WMAI (2015). *Code of Practice for Internal Wet Area Membranes (Selection, design, installation)* (2nd ed.). New Zealand: Waterproofing Membrane Association (NZ) Incorporated.

[20] Restroom Association (2013). *A Guide to Better Public Toilet Design and Maintenance* (3rd ed.). Singapore: Restroom Association.

[21] Public Utilities Board (2016). Plumbing Works. Water Supply. Retrieved on March 22 from https://www.pub.gov.sg/watersupply/plumbingworks

[22] Building and Construction Authority (2014). *Prefabricated Bathroom Unit (PBU)*. Singapore: Building and Construction Authority (BCA).

[23] Public Utilities Board (2004). *Code of Practice on Sewerage and Sanitary Works* (1st Adden.). Singapore: Public Utilities Board (PUB).

Normative References/Standards Referred to for Wet Area

- ANSI A118.15:2012 — American National Standard Specifications for Improved Modified Dry-Set Cement Mortar
- AS 3740:2010 — Waterproofing of domestic wet areas (Standards Australia)
- ASTM C1400-11 — Standard Guide for Reduction of Efflorescence Potential in New Masonry Walls
- BS 5385-1:2009 — Wall and floor tiling. Design and installation of ceramic, natural stone and mosaic wall tiling in normal internal conditions. Code of practice
- BS 5385-3:2014 — Wall and floor tiling. Design and installation of internal and external ceramic and mosaic floor tiling in normal conditions. Code of practice
- BS 6093:2006+A1:2013 — Design of joints and jointing in building construction. Guide
- BS 6150:2006+A1:2014 — Painting of buildings. Code of practice
- BS 6270-3:1991 — Code of practice for cleaning and surface repair of buildings. Metals (cleaning only)
- BS 8000-0:2014 — Workmanship on construction sites. Introduction and general principles
- BS 8000-11:2011 Workmanship on building sites. Internal and external wall and floor tiling. Ceramic and agglomerated stone tiles, natural stone and terrazzo tiles and slabs, and mosaics. Code of practice
- BS 8000-9:2003 — Workmanship on building sites. Cementitious levelling screeds and wearing screeds. Code of practice
- BS 812-109:1990 — Testing aggregates. Methods for determination of moisture content
- BS 8204-1:2003+A1:2009 — Screeds, bases and in situ floorings. Concrete bases and cementitious levelling screeds to receive floorings. Code of practice
- BS 8204-2:2003+A2:2011 — Screeds, bases and in situ floorings. Concrete wearing surfaces. Code of practice
- BS 8210:2012 — Guide to facilities maintenance management
- BS 8215:1991 — Code of practice for design and installation of damp-proof courses in masonry construction

- BS 8221-1:2012 — Code of practice for cleaning and surface repair of buildings. Cleaning of natural stone, brick, terracotta and concrete
- BS 8221-2:2000 — Code of practice for cleaning and surface repair of buildings. Surface repair of natural stones, brick and terracotta
- BS 8542:2011 — Calculating domestic water consumption in non-domestic buildings. Code of practice
- BS 8554:2015 — Code of practice for the sampling and monitoring of hot and cold water services in buildings
- BS 8595:2013 — Code of practice for the selection of water reuse systems
- BS EN 12004-1:2017 — Adhesives for ceramic tiles. Requirements, assessment and verification of constancy of performance, classification and marking
- BS EN 12056-2:2000 — Gravity drainage systems inside buildings. Sanitary pipework, layout and calculation
- BS EN 1504-5:2013 — Products and systems for the protection and repair of concrete structures. Definitions, requirements, quality control and evaluation of conformity. Concrete injection
- BS EN 1992-1-1:2004+A1:2014 — Eurocode 2: Design of concrete structures. General rules and rules for buildings
- BS EN 1992-3:2006 — Eurocode 2. Design of concrete structures. Liquid retaining and containing structures
- BS EN 771-1:2011+A1:2015 — Specification for masonry units. Clay masonry units
- BS EN ISO 12944-4:1998 — Paints and varnishes. Corrosion protection of steel structures by protective paint systems. Types of surface and surface preparation
- BS EN ISO 12944 series — Paints and varnishes. Corrosion protection of steel structures by protective paint systems
- BS EN ISO 1513:2010 — Paints and varnishes. Examination and preparation of test samples
- BS ISO 10880:2017 — Non-destructive testing. Infrared thermographic testing. General principles
- ISO 10545-1:2014 — Ceramic tiles — Part 1: Sampling and basis for acceptance
- ISO 13007-1:2014 — Ceramic tiles — Grouts and adhesives — Part 1: Terms, definitions and specifications for adhesives
- ISO 13007-5:2015 — Ceramic tiles — Grouts and adhesives — Part 5: Requirements, test methods, evaluation of conformity, classification and designation of liquid-applied waterproofing membranes for use beneath ceramic tiling bonded with adhesives
- ISO 28841:2013 — Guidelines for simplified seismic assessment and rehabilitation of concrete buildings
- ISO/DIS 13006 — Ceramic tiles — Definitions, classification, characteristics and marking
- ISO/NP 4356 — Bases for the design of structures — Deformations of buildings at the serviceability limit states

- ISO/TS 7024:2005 — Plastics piping systems for soil and waste discharge (low and high temperature) inside buildings — Thermoplastics — Recommended practice for installation
- SS 509-1:2015 — Code of practice for cleaning and surface repair of buildings — Part 1: Cleaning of natural stone, brick, terracotta, concrete and rendered finishes
- SS 509-2:2005(2015) — Code of practice for cleaning and surface repair of buildings — Surface repair of natural stones, brick, terracotta and rendered finishes
- SS 542:2008 — Code of practice for painting of buildings
- SS CP 68:1997 — Code of practice for ceramic wall and floor tiling
- SS CP 82:1999 — Code of practice for waterproofing of reinforced concrete buildings
- SS EN 1992-1-1:2008 — Eurocode 2: Design of concrete structures, Part 1-1 General rules and rules for buildings
- SS EN 934-2:2015 — Admixtures for concrete, mortar and grout — Part 2: Definitions, requirements — Concrete admixtures — Definitions, requirements, conformity, marking and labelling

Chapter 3

Façades

Introduction

Façades are the interface between external and interior environments, and have an immense impact on climate control/regulation, the level of daylight entering a building, energy consumption and the building's overall carbon emission intensity.

A façade can be traditional (e.g. masonry) or modern (e.g. cladding, prefabrication, curtain wall, etc.), depending on the building's spatial, aesthetic, energy performance, heat transfer, lighting, sound and comfort requirements. Finishes should be compatible with the wall system, as they are the façade's outer skin, which provides both aesthetics and protection against weathering. Traditional façade systems are mainly made of masonry and/or reinforced concrete and finished with cement render and painted, or finished with various cladding materials. Modern façade systems include large panel curtain walls or green features such as photovoltaic cells, vertical greening, LED, automatic blinds/louvers, high performance glazing, and nano coatings [1].

The common defects on façades include: (a) cracking; (b) joint/sealant failures; (c) efflorescence; (d) rising damp/water penetration; (e) abrasion; (f) corrosion and (g) delamination. Inaccessibility, wind driven rain, and falling hazards are also major issues to consider in the design, construction and maintenance of façades.

Adherence to proper design, construction and maintenance practices would safeguard the long term maintainability of the façade and the entire building. A list of relevant design, construction and maintenance guidelines for façades are highlighted in the succeeding tables. Most of the identified structural defects pertain to cracks, alkali silica reaction, movement of joints, rising dampness, corrosion and falling hazards (e.g. mechanical fixtures, window panels, detached tiles, etc.). The

View of building façades in Singapore.

architectural defects refer to the inadequate knowledge on the durability of the materials selected (e.g. cracking of glass cladding, curtain wall staining, buckling of metal cladding), sealant deterioration, delamination, weather-tightness (airtightness and water-tightness), window fenestration (leakage rate of windows) and façade staining. Additional relevant façade guidelines, such as glare mitigation and shading/ weather control devices, as well as façade cleaning methods, are provided. Façade defects such as façade access, and issues with fixtures and fittings are dealt with in the service component tables.

The proper design of a façade should take weather-tightness, water runoff from the façade, and ease of cleaning into consideration. Defects due to water penetration through jointing, water/air ingress and façade staining can be addressed by adhering to proper window and joint design detailing requirements by various standards and codes. The failure of joint sealants can also be avoided by providing careful joint design and detailing, choosing the right sealant, ensuring the proper surface preparation and performing quality workmanship during installation. Adequate provision of expansion joints to accommodate differential thermal/moisture movements, and the proper selection, construction and maintenance of sealant will ensure surface continuity and weather-tightness [2].

Any potentially unsafe conditions of building façades can jeopardise public safety and surrounding properties if they remain unaddressed. Falling hazards can be avoided with sufficient knowledge on the durability of fixings, quality workmanship during installation, and proper surface preparation and adhesions. Façade ordinances help cities ensure public safety, prevent injury and death, and avoid property damage by imposing periodic inspections of building façades [3, 4].

Accessibility for maximum coverage of external wall areas for effective maintenance requires the design conceptualisation of the building's overall form right from the planning/design stage. It should be done with adequate provisions given for future operation of its access systems [5].

3.1 Structural

The structural components of a façade require adequate provisions in terms of design, construction and maintenance, to prevent premature deterioration and to achieve optimum performance under specific external environmental exposure.

Problem	Design	Construction	Maintenance
Cracks 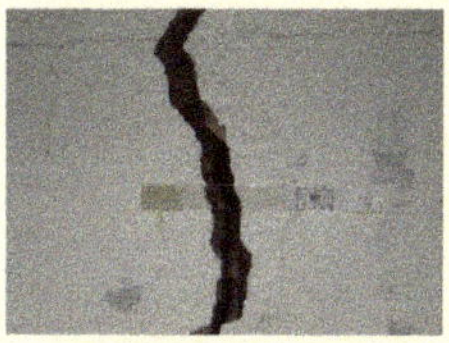 Crack on concrete  Crack on tiled wall	Tall slender concrete structures should be designed with due consideration of the effects of lateral deflection, and be within acceptable vibration limits (BS EN 1992-1-1: 2004+A1:2014, SS EN 1992-1-1:2008, SS EN 1992-1-2:2008 (2015), SS CP 65-1:1999). Any deflection/ deformation of the concrete structure due to vertical loading should be compatible to the degree of movement acceptable by other elements (i.e. the finishes, services, partition, glazing, and cladding) (BS EN 1992-1-1:2004+A1:2014).	Use two-stage joints for precast façade construction to ensure higher water-tightness performance, since doing so avoids seepage through hairline cracks — as is the case with one-stage joints. Horizontal joints for load bearing walls should be sealed off with non-shrink grout. Minimise cracks in rendered brick walls by using appropriate mix ratio, thickness and number of coats.	Decisions for surface repairs should consider the ease of access for future work, relative cost of hiring and erecting scaffolding, and the probable frequency of maintenance (BS 8221-2:2000, SS 509-2:2005). Record and retain documentation of all executed works on façades, including photographs and non-destructive survey techniques, to provide background information prior to further assessment or work.[1]

(*Continued*)

[1]See *Maintainability of Buildings, Maintenance. Available at: http://www.hpbc.bdg.nus.edu.sg/?page_id=8804 (accessed May 12, 2017)* for more information about initiating a thorough inspection and maintenance programme.

(*Continued*)

Problem	Design	Construction	Maintenance
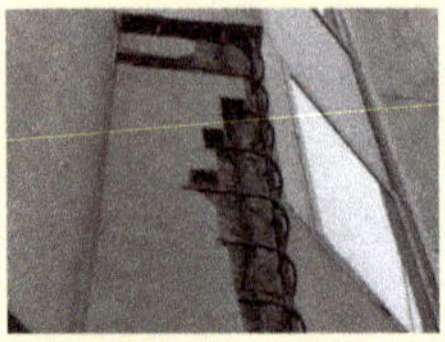 Crack on cladding  Crack on brick  Crack at corbel supporting a full precast beam  Crack on façade wall  Crack on façade wall	Recommend Limit States Method for design and verification of structure for durability of façade structures. Refer to modelling of deterioration process (ISO 13823:2008). Understand the causes, effects, and methods of prevention and repair for cracks in precast concrete wall panels (BS EN 13369:2013, SS CP 81:1999). Limit the design crack width with reference to SS CP 65-2:1996(1999). Conduct laboratory mechanical tests to measure deformations on horizontal joints between load-bearing walls and concrete floors (ISO 7845:1985). Design of curtain wall systems as per CWCT (*Curtain Wall Installation Handbook* [7]) to maintain structural integrity (BS EN 1991-1-1:2002, BS EN 1991-1-4:2005+A1:2010, ASTM E1300 – 16, BS 8213-4:2016, BS 4873:2016, BS EN 1999-1-1:2007+A2:2013, BS EN 1999-1-4:2007+A1:2011, BS 1161:1977, BS EN 1090-2:2008+A1:2011, BS EN 1993-1-1:2005+A1:2014).	Provide bonding bars at interfaces between different material in order to minimise cracks (e.g. where brick wall abuts concrete). Alternatively, the bonding bars can be cast together with the concrete member [6]. At the completion of the construction stage, minor repair work or fixing adjustments may be acceptable. Enhance the durability of vulnerable parts of construction; ensure that surfaces exposed to water are freely drained; provide adequate cover to steel; use protective coatings for either steel or the concrete, or both (BS EN 1992-1-1:2004+A1:2014, SS EN 1992-1-1:2008, SS EN 1992-1-2:2008 (2015), SS CP 65-1:1999). Components whose predicted service life is less than the design life of the structure must be inspectable and replaceable (ISO 13823:2008).	Conduct semi-annual inspections of stone-wall elements, inspecting all elevations. Keep accurate and cumulative records of inspection findings (ASTM C1496-11, BS 8298-1:2010). Regular cleaning is critical for the long-term durability and appearance of natural stone façades. Perform periodic joint repairs, (i.e. sealant replacement, tuck pointing, and cleaning) (ASTM C1496-11, BS 8298-1:2010). Perform repairs and restoration works as per guide ASTM C1722-11, BS 8221-2:2000. For cases of cracked or broken stones: (1) seek assistance for stone replacement; (2) if fragments are stable and secure, tuck point or caulk crack with sealant; (3) monitor closely for additional cracking or movement. Monitor joints for continued movement (ASTM C1496-11, BS 8298-1:2010).[2]

(*Continued*)

[2]See *Maintainability of Buildings, Concrete Cracks. Available at: http://www.hpbc.bdg.nus.edu.sg/?page_id=8307 (accessed May 12, 2017)* for details on concrete crack classifications and concrete durability.

(*Continued*)

Problem	Design	Construction	Maintenance
Alkali-Silica reaction (ASR) 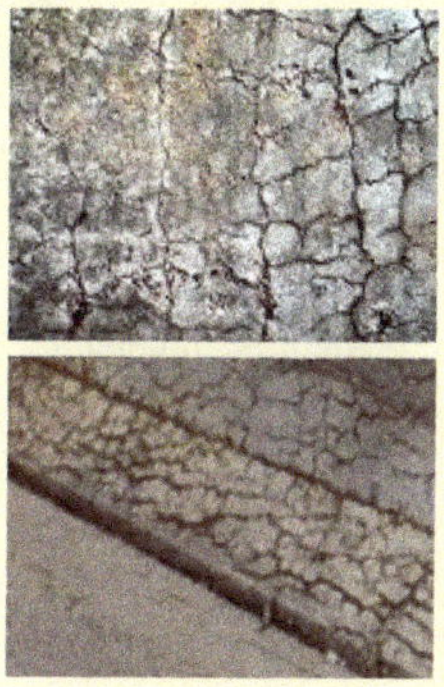 Alkali-silica reactions in concrete	Carefully analyse all cementitious material and aggregates during material selection and sourcing. Specify the use of non-reactive aggregates. Recommend low alkali cement, and take steps to prevent alkaline solutions from coming into contact with and penetrating the concrete [8]. Test aggregate supply to determine potential reactivity (ASTM C227 — 10, ASTM C1260 — 14, ASTM C295/C295M — 12, BS EN 932-5:2012, BS 812-123:1999).	Use information from field performance history to determine the susceptibility of ASR. Perform testing for ASR in cementitious materials and aggregate in mortar bars as per guides ASTM C1567-13/ASTM C1293-08b (2015). If historical experience or test results show a potential concern, provide additional supplementary cementitious materials (SCMs) to inhibit ASR (ASTM C618 — 15, ASTM C989/ C989M — 16e1).	Where there is potential for ASR, be on the lookout for typical visual symptoms such as: unusual expansion of concrete, evidenced by longitudinal cracks; map cracking (random cracking pattern); closed joints; spalled surfaces; displacement of adjacent structural components; pop-outs; efflorescence (surface deposits); or discolouration (dark or blotchy areas). Identify ASR using a petrographic microscope (ASTM C856 — 17), electron microscopy, or an ASR detect kit (coloured dye field test kit).
Movement joints  Shrinkage crack at façade movement joints	Use suitable joints to accommodate movement as per BS EN 13830:2015, SS CP 96:2002 (2011) for sealant classification and selection criteria. For joint design in precast concrete refer to BS 6093:2006+ A1:2013, BS EN 14992:2007+ A1:2012, SS CP 81:1999. Specify two-stage joint for joints of external façade and walls to ensure durable waterproofing (BS 6213:2000+A1:2010, SS CP 82:1999).	Conform to the requirements in forming movement joints (BS EN 1992-1-1:2004+ A1:2014, BS 6093: 2006+A1:2013, SS CP 65-1:1999) Joint should prevent movement, spread of flame, transmission of airborne sound between dwellings and be weathertight (if external) (BS EN 13369:2013, SS CP 81:1999). Use overlapping to ensure water-tightness even under vertical movements [6].	A visual inspection of the façade is key in identifying defects in movement joints. Inspections for general defects and moisture ingress can be carried out either quarterly or during façade cleaning exercises. Recommend housekeeping of joints by cleaning on a routine basis to remove any dirt or debris that may inhibit their movement. Repair defects at movement joints through proper re-application of sealants.[3]

(*Continued*)

[3]See *Maintainability of Buildings, Sealants. Available at: http://www.hpbc.bdg.nus.edu.sg/?page_id=4170 (accessed May 12, 2017)* for further information of sealant types and their respective properties and durability; and details on cleaning, maintenance, defects and, repairs.

(*Continued*)

Problem	Design	Construction	Maintenance
Rising dampness 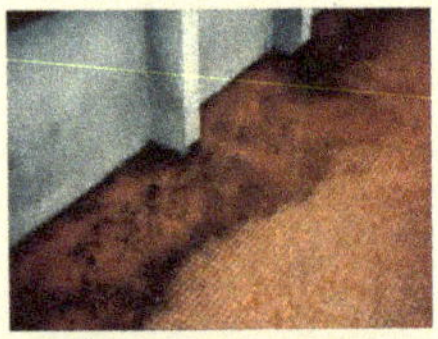  Rising dampness	Conform to the proper waterproofing design detailing for reinforced concrete structures. To avoid rising dampness, use suitable DPM/ DPC for the site ground conditions (SS CP 82:1999) Concrete façade design needs to take into consideration a sound understanding of the exposure conditions on site (BS EN 1992-1-1:2004+A1:2014). Specify a dense concrete with minimum water to cement ratio to reduce permeability as permeability is closely related to concrete durability.	Provide adequate damp-proof course/membrane at a height of at least 150 mm above the surrounding finished floor level, to prevent upward movement of moisture through capillary action or rainwater bouncing off the ground. Provide adequate surface drainage and adequate coating, and/or hydrophobic materials, and/or chemical injection as moisture barrier (BS 8215:1991, BS6576:2005).	Identify tell-tale signs of moisture entry/rising dampness (e.g. wetness, staining, darkening due to trapped moisture, discolouration, efflorescence deposits). Diagnose rising dampness through surface-breaking flaws with a liquid penetration test.[4] To remedy rising damp, expose lower surfaces of façade and allow drying; then damp-proof and provide additional drainage (ASTM C1496-11, BS 8298-1:2010).
Corrosion of RC  Spalling of concrete due to corrosion of rebars	Adhere to the general reinforced concrete requirements (ISO 15673:2016). Conform to the general rules for designing concrete and steel composite structures (BS EN 1994-1-1:2004, SS EN 1994-1-2:2009).	Ensure reinforcements are surrounded with adequate thickness of good quality, well-compacted, homogeneous concrete, free from honeycombing or other defects. Perform material selection with the aim to reduce chloride content in concrete in mind, so as to reduce corrosion risk of embedded metal.	Conduct appropriate maintenance of the fenestration product and its interfaces with the wall system to ensure long-term delivery of the desired water penetration resistance. Maintain records of building use, maintenance and performance problems, as well as responses to those problems (ASTM E241-09(2014)e1).

[4]See *Maintainability of Buildings, Liquid Penetration Test. Available at: http://www.hpbc.bdg.nus.edu.sg/?page_id=112 (accessed May 12, 2017)* for more details on the principle, the test procedures, application, and advantages and limitations of the test.

FALLING OBJECTS FROM FAÇADES

Examples of falling façade materials/components/features

Life threatening falling objects (including materials, components or features) from building façades have been reported globally. It is imperative that all façades must be designed, constructed and maintained to ensure public safety. To this extent, periodic inspections and repairs on façades must be conducted. Aside from the **ASTM Guide Book STP 1444: Building Façade Maintenance, Repair and Inspection**, other relevant standards that cover methods and procedures for inspection, evaluation, and reporting for the periodic inspection of building façades for unsafe conditions include:

- **ASTM E2270-14 — Standard Practice for Periodic Inspection of Building Façades for Unsafe Conditions:** This standard practice establishes the minimum requirements for conducting periodic inspections of building façades to identify unsafe conditions that could cause harm to persons and property. It addresses the required content of the façade inspection report, in order to convey to the specifying authority, the condition of the façade, and to allow for comparisons against the façade's condition at other times.
- **ASTM E2841-11 — Standard Guide for Conducting Inspections of Building Façades for Unsafe Conditions:** This guide establishes the procedures and methodologies for conducting inspections of building façades — including those that meet the inspection criteria/comply with Practice E2270.

As in the case of falling window panels from a building, reference to the following design, construction and maintenance guidelines may prevent the occurrence of said defect:

- **Design:** Comply with the requirements for curtain wall construction (BS EN 13830:2015, SS CP 96:2002(2011)). Refer to SS 212:2007 (BS 4873:2016) for the minimum performance of aluminium alloy windows.
- **Construction:** Provide maintenance and replacement requirements for structural sealant glazing systems (ASTM C1401-14, BS EN 13022-1:2014). Use stainless steel ironmongery in areas with a high risk of corrosion and degradation.
- **Maintenance:** Conduct continual maintenance of the exterior seals to maintain a weather-tight condition (ASTM C1401-14, BS EN 13022-1:2014). Routinely check aluminium rivets, screws, hinges, sliding shoes and tracks. Lubricate bi-annually.

3.2 Architectural

The façade's architectural components contribute to its aesthetic value and durability. The design, construction and maintenance specifications for the façade's architectural components bring the opportunity for optimal façade performance.

Problem	Design	Construction	Maintenance
Material selection and handling  Staining of curtain wall  Cracking of glass cladding  Buckling of metal cladding  Crack on cladding	Design curtain walls to withstand live loads resulting from regular maintenance activities (BS EN 13830:2015, SS CP 96:2002 (2011)). Recommend materials for curtain wall elements as per BS EN 13830:2015, SS CP 96:2002(2011). Classify clay and calcium silicate brickwork and select the appropriate maintenance method (BS 8221-1:2012, SS 509-1:2015).[5] Façade hardware and other components such as mullions, panels, fascia, column covers, windows, doors, trim, roofing, gutters, and flashing should be made of corrosion-resistant material. If different metallic materials are used, designers must ensure that the materials are compatible and able to avoid galvanic corrosion and the like, in incidents of both contact and run-off (ISO 15686-1:2011).	Strict supervision required to achieve quality for surface evenness, finishing, and alignment without noticeable staining or cracking. Refer to relevant façade material standards as per CONQUAS [9]. Precast walling components should be transported, handled and stored so as to avoid damages. Handle material as per guidelines set out in SS CP 81:1999. (See also BS EN 13369:2013). Components of structural timberwork should be fabricated, stored and handled on site as per BS EN 1995-1-1:2004+ A2:2014, SS CP 7:1997 (2014). Facility manager and designers to make maintenance manual during commissioning (including maintenance life and other construction requirements) (ASTM E2266-11).	Inspect and test adhesion of structural/weather seal sealants. Check for movement failures, moisture, condensation, and on the condition of the organic coatings on metal surfaces (ASTM C1401-14). All buildings with more than 25% of façade in curtain wall must be inspected every 6 to 12 months for deterioration of aluminium framing. Condition of coating, hardware, glass, sealants and weather seals must also be checked on (SS CP 96:2002(2011), Refer also to BS EN 13830:2015). Proper maintenance should be carried out on structural timberwork components to maintain effectivity during the intended lifespan (BS 7543; BS EN 1995-1-1:2004+A2: 2014, SS CP 7:1997 (2014)).

(*Continued*)

[5]See *Maintainability of Buildings, Façade Assessor. Available at: http://www.hpbc.bdg.nus.edu.sg/?page_id=360 (accessed May 12, 2017)* for a useful tool to calculate façade durability and estimated cleaning/replacement costs; based on the material and site location.

(*Continued*)

Problem	Design	Construction	Maintenance
Sealant deterioration  Sealant deterioration on façade 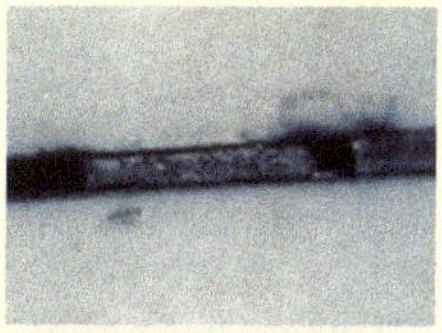  Loss of cohesion in sealent	Select sealant joint type and mode of application as per BS 6213: 2000+A1:2010, SS CP 96:2002 (2011); including requirements for quality management programme. Conform to the guidelines on the selection of construction sealants as per BS 8221-2: 2000. (See also SS 509-2: 2005 (2015)). Comply with the grading of sealant (i.e. pourable or non-sag) as per ASTM C920 — 14a (tested in accordance with ASTM C639 — 15). Make provisions for access for regular sealant inspection and avoid placing design features/services across joints that impede access for maintenance [6].	Application of sealants to follow recommended guidelines as per BS 8000-0:2014, BS 6093:2006+A1:2013, SS CP 96:2002 (2011). Perform dynamic peel test and dynamic tensile test on structural sealants (ISO 28278-2:2010). Conduct non-destructive and destructive inspection procedures of weatherproofing sealant joints as per guide ASTM C1521-13. (See also BS EN 15651-1:2017).	Multi-story structures require a periodic façade inspection at an interval of about 5 years to identify areas where remedial sealant repair or maintenance work is required (ASTM C1193-16). Replace failed sealants immediately (ASTM C1401-14, BS EN 13022-1:2014). Suspected failure of waterproofing to be tested using ASTM C1521-13 (see also BS EN 15651-1:2017). Remedial work for glazing sealant should be in accordance with ASTM C1487-02(2012).
Corrosion of metal cladding  Dirt stains and streaks of corrosion on metal cladding	Material selection for metal cladding should be based on aesthetics, cost, availability, formability and corrosion resistance (BS 5427:2016). Evaluation of corrosion resistance of the metal should be based on product warranty, environmental	Ensure the use of manufacturer's fasteners/ brackets/stiffeners and all fixings and accessories to prevent sacrificial corrosion, as well as transmit all imposed loads and stresses of the cladding. Avoid scratching and damage of protective	Conduct inspections to look out for signs of deterioration in aluminium framing (BS EN 13830:2015, SS CP 96:2002(2011)). Conduct routine cleaning to remove surface contaminants from metals in order to ensure maximum corrosion resistance of

(*Continued*)

(Continued)

Problem	Design	Construction	Maintenance
 Delamination and corrosion of metal cladding  Corrosion of metal cladding	effects (external, internal, industrial, and acid rain), and maintenance (unwashed areas). Aluminium infill panels to comply with BS EN 485-1:2016. Mild steel infill panels to comply with BS EN 10346:2015.	coating/film during site handling and installation. Alert designer/supplier if cladding is exposed to high concentrations of sulphur and chloride containing gases. Refer to lightning protection as per SS 555: Parts 1 to 4 : 2010. Remove drill swarf and other visible contaminants from the cladding surface to avoid corrosion [10].	the metal. The cleaning process chosen should be selected based on type of contaminant, the required degree of cleanliness, and cost (ASTM A380/A380M-13, BS EN 1993-1-4:2006+A1:2015, BS EN ISO 12944-8:1998). Special care should be taken in order not to damage the cladding surface during cleaning/maintenance [10].

SHADING AND WEATHER CONTROL DEVICES

Sun shading devices (Marina Bay Sands, Singapore)

Design: Calculate the effects of shading devices on the thermal/optical properties of a window system as per ISO 15099:2003. Provide weather protection features (e.g. drip, water bay, canopy, coping, etc.) for weather protection of windows or doors on external walls.

Construction: Ensure consistency of functional, endurance and safety performance for shading devices as per BS EN 13561:2015. For external blinds and shutters, also refer to BS EN 1932:2013. Ensure external shading device performance as per BS EN 13561:2015. (See also BS EN 1932:2013 for required wind resistance criteria).

Maintenance: Test methods of loading for shading devices (BS EN 1932:2013). Provide safe access provisions for maintenance workers to reach external shading and weather control devices (BS 8221-1:2012, SS 509-1:2015).

Problem	Design	Construction	Maintenance
Delamination of façade  Spalling, paint peeling and plaster wall delamination 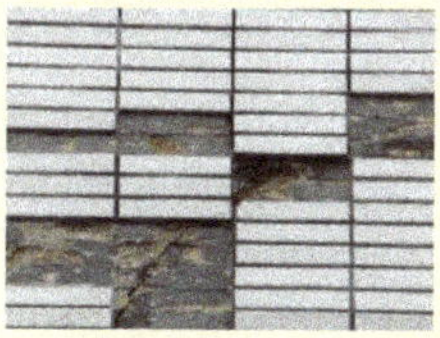 Massive tile cladding delamination[6]  Façade wall delamination	Recommend surface treatment and protection as per SS 509-2: 2005(2015). (See also BS 8221-2:2000). Ensure proper detailing of window, door, and abutment points where water seepage can occur and cause delamination. Design movement joints as required on the structural concrete (BS 6093:2006+A1: 2013). Impede water ingress to avoid tile delamination by detailing metal capping or flashing to protect points of weakness of exposed tile edges at window, door or abutment. Horizontally tiled surfaces should be avoided and openings should be set back within the elevations.	Sufficiently remove curing agent (used for early stripping of formwork) to ensure proper adhesion on the substrate. Ensure proper substrate preparation (cleaning) prior to application of plaster/tile finishing. Perform quality workmanship on façade through the correct handling and angle of application to prevent delamination (BS 8000-0:2014). Adhere to the performance requirements for curtain walling, including safety as per BS EN 12179:2000. (See also ASTM E2270-14).	Repair of spalled surfaces should be done by cutting loose and flaking material to the base and replacing with new bricks or blocks, or made with layers of mortar. Consolidate weathered masonry (to stabilise the degradation) as per SS 509-2:2005(2015). (See also BS 8221-2:2000). Maintenance and repair of renders should be carried out in accordance with BS EN 13914-1:2016. (See also ASTM C926 — 17). Attend to cracks on façade surface promptly to control moisture ingress in order to avoid delamination. Re-tile debonded tiles or over-clad existing façade to remedy delamination [11].

(*Continued*)

[6]Avoid tile delamination by using ceramic tiles with suitable water absorption and thermal expansion properties on external façade. Further, the tile bed has to be sufficiently flexible to cope with the moisture and thermal movements. Cement based polymer latex or resin are suitable for tiles. The grouting material is normally of the same type as tile bed, compatible to the tile bed and the tile. Single-fired vitrified wall tiles (glazed or unglazed) are especially suited for external walls subject to thermal shock conditions on tropical climates. The tiles shall be fully tested to meet the performance requirements stipulated in the contract before mass installation (AS 3958.1-2007/AMDT1-2010; SS CP 68:1997).

Problem	Design	Construction	Maintenance
Weather-tightness 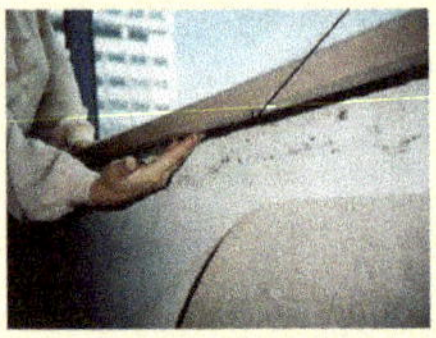 Gap underneath window sill allows outside air to leak into the interior  Water marks on window frame  Water seepage onto window sill	Specify and ensure joint sealant performance are of suitable form to withstand air penetration during assembly, transportation, installation and operation of the curtain wall system (BS EN 12152:2002, BS EN 13830:2015, SS CP 96: 2002(2011)). Comply with the recommendations for water-tightness of external walls as per SS CP 82:1999. Refer to the provisions for water-tightness of precast concrete slab and wall panels as per BS EN 13369:2013, SS CP 81:1999. Conform to the preventive measures for water leakage and weatherproof joints (BS EN 12154:2000, BS EN 13830:2015, SS CP 96:2002 (2011)).	Air permeability of joints between precast concrete external wall components is measured in accordance with ISO 6589:1983 or BS EN 12153:2000. Test weather resistance of the external façade against acceptable performance criteria.[7] Detect water leakage through façade by simulating rainwater penetrations under pressure using a water-tightness test.[8] Maintain water-tightness with respect to rainwater which would otherwise give rise to moisture stains on internal face, or cause damage to the façade or other building elements (ISO 7361:1986). Water-tightness of joints between two prefabricated ordinary concrete external wall components (ISO 7729:1985).	Integral seals in window units require maintenance or replacement within the duration of the service life (ASTM E2266-11). Check condition of windows and caulk seals annually (ASTM E241-09(2014)e1). Maintain repair records to identify a pattern of leakage; and to identify if repairs may be causing or contributing to current leakage. Use of maintenance records to diagnose buildings with chronic leakage problems (e.g. areas that have been subjected to several attempts at remediation). Suspected water leakage in glazing systems should be evaluated using Guide ASTM E2128-12. NDT methods can be used (e.g. thermography, fiberscope, elastic recovery meter) (Refer also to BS EN 1026:2016).

(*Continued*)

[7]See *Maintainability of Buildings, Weather Performance Test. Available at: http://www.hpbc.bdg.nus.edu.sg/?page_id=113 (accessed May 12, 2017)* for detailed instructions on the applications, advantages, limitations and test procedure of weather performance testing.

[8]See *Maintainability of Buildings, Water Tightness Test. Available at: http://www.hpbc.bdg.nus.edu.sg/?page_id=114 (accessed May 12, 2017)* for further details of water-tightness test, which involves a chamber mounted onto a vertical element to detect leakage or dampness.

(*Continued*)

Problem	Design	Construction	Maintenance
Window/fenestration[9]  Water seepage onto window sill 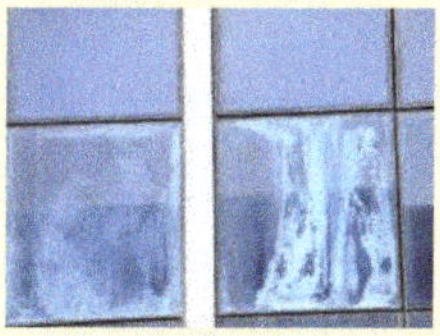 Window film defect 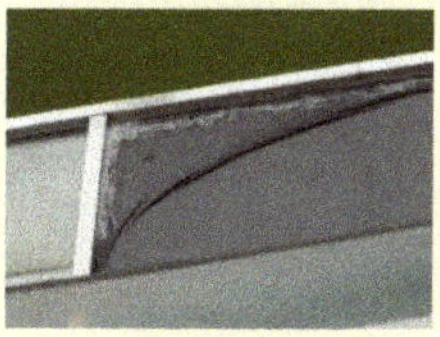 Film delamination	Conform to the guidelines for calculations of thermal performance of windows, doors and shading devices as per ISO 15099:2003, BS EN ISO 12631:2012. Comply with the required thermal performance of a curtain wall (overall U value) (BS EN ISO 12631: 2012, BS EN 13830:2015, SS CP 96: 2002(2011)). Water penetration must be prevented by means of adopting good practices in sealant choice and application. Choice of suitable sealant is especially important for window performance in the tropics. Ensure compatibility of sealants with one another. Polysulphide sealant has good movement capacity	Keep an attic stock on-site for future uncertainties, especially important for reflective or low-emissivity coated glass as replacement stock may result in colour or reflectivity matching problem (ASTM C1401-14). Test method for weather-tightness (air leakage and water tightness) of aluminium alloy windows (BS 4873: 2016, SS 212: 2007). Cleaning of fenestration products must be in strict accordance with the manufacturer's installation instructions. Replace all failed sealant with better quality, UV-resistant sealant for use in local climate so as to accommodate the movement joints.	Conduct regular maintenance of the fenestration product and its interfaces with the wall system to ensure water penetration resistance (ASTM E241-09 (2014)e1). Clean glazing system's exterior surface to control accumulation of environmental pollutants, as well as to avoid staining and disfiguration of glass. Conduct periodic maintenance as required for window gasket seals and operating hardware (ASTM C1401-14). Use cleaning solvents in strict accordance with solvent manufacturer's instructions and applicable codes, safety regulations, and environmental regulations. MEK (Methyl ethyl ketone) and similar solvents may damage organic sealants, gaskets, and finishes used on fenestration products (ASTM E2112-07 (2016), BS 8213-4:2016.

(*Continued*)

[9]See *Maintainability of Buildings, Double Glazed Units. Available at: http://www.hpbc.bdg.nus.edu.sg/?page_id=4194 (accessed May 12, 2017)* for more on maintenance and cleaning of double glazed units and information on common defects and repairs.

(Continued)

Problem	Design	Construction	Maintenance
 Water penetration from façade due to sealant failure	and high chemical resistance, is suitable for submerged application and has a long service life. Other generic sealants, such as silicone, are also acceptable.	Maintenance requirements should be passed on during commissioning stages.	Refer also to BS EN 1027:2016 for testing methods). Avoid installing new sealant over an existing one during maintenance operation.

GLARE MITIGATION

Contemporary buildings mostly use glass and metal façades and metal roofs, which in most cases cause sunlight to be reflected off these building materials. Daylight reflectance is regulated in view of the potential disamenity associated with glare from buildings. Daylight reflectance (which is the sum of specular reflectance and diffuse reflectance) is the measurement of the percentage of visible light that is reflected off a material [12].

Reflectivity of glass causes glare to the surrounding environment

Design considerations:

- Ensure that the reflectivity of any material (other than glass) used in façades, have a specular reflectance not exceeding 10%, without any control on daylight reflectance.
- As for glass façades/roofs, it is allowable to use of any material with daylight reflectance not exceeding 20% [12]. For curtain walling, refer also to Building Control Regulations (BS EN 13830:2015, SS CP 96:2002(2011)).
- For testing standard for reflectance values, BCA recommends that the test be conducted in accordance with ASTM E903 [12].

Problem	Design	Construction	Maintenance
Staining  Algae growth on masonry  Efflorescence and dirt staining  Dirt streaking  Staining on Brick wall  Rust and dirt stains on façade	Specify material and application methods for water repellency of porous masonry (BS 8221-2:2000, SS 509-2:2005(2015)). Throw off water from the façade altogether through an outward projecting sill or overhanging eaves (which incorporate a throat or drip lines on its underside) or provide blocking features such as copings/flashings. Use efficient scupper drains/downpipes to channel water down and away from the façade. Specify paint system which is permeable to avoid any paint defects which may cause staining. Render the detailing for open joints as opposed to butt joints to avoid sealant staining. Recommend joint designs which are able to retain runoff within joints and expansion joints designated to provide vertical runoff carrying dirt down along the façade surface [12].	Façade surfaces should be painted evenly with no patchiness. The finished texture should be uniform in colour [8]. Ensure proper rendering to control surface granularity and local faults as it influences colour uniformity of the external façade (ISO 7361:1986). Correct sealant applications to ensure consistent and continuous quality. Avoid misaligned panels of cladding (BS 8000-0:2014). Refer to the planning of painting programme, including inspection regime (initial and routine inspections) for buildings as per BS 6150:2006+A1:2014, SS 542:2008. Use self-cleaning coatings on newly built substrates for increased success in its performance. All construction/repair works on a façade surface must be done prior to application of a water repellent (BS 8221-2:2000, SS 509-2:2005(2015)).	Consider availability of adequate water supply, drainage provisions and electrical power supply to choose façade cleaning method. Records of cleaning operations (including; photographs before and after cleaning, and drawings of nature of deposits, thickness and patterns) should be kept for buildings of significance (BS 8221-1:2012, SS 509-1:2015). Maintain façade in a state as near as possible to its new condition. Ease of façade maintenance can be expressed by frequency of necessary maintenance operations; labour and supplies necessary for each maintenance operation; and notice of possible ways of removing stains, graffiti, etc. (ISO 7361:1986). Detect/determine staining of porous substrates by joint sealants (ISO/NP 16938-1). Comply with the recommendations for treatments for controlling organic growth (BS 8221-2:2000, SS509-2:2005 (2015)).

(*Continued*)

(*Continued*)

Problem	Design	Construction	Maintenance
 Sealant staining on glass surface façade	Specify façade self-cleaning applications (e.g. TiO_2, superhydrophobic paint products, etc.) with due consideration given to site orientation, sunshades and protruding features.	Serviceability of exterior façade surfaces is important as it dictates the building's individual and corporate identity (ASTM E1667-95a(2012)).	Repair painted surfaces damaged by wear and tear; wash down; remove defective paint film; apply sealer/primer (if necessary); and repaint. (BS 6150: 2006+A1:2014, SS 542:2008).

SOME EXAMPLES OF FAÇADE CLEANING METHODS [13]

Brick wall cleaning

Brick masonry

Brick walls can be cleaned with chemicals in conjunction with water rinsing. Dirt stains and biological growth could be water blasted to remove them from the brick wall. Efflorescence has to be brushed or scraped off from the surface whenever the salts appear. Acidic cleaners containing dilute mineral acids such as hydrochloric, hydrofluoric, phosphoric and/or organic acids such as acetic and citric acids are used to remove heavy soiling from most brick masonry walls.

Cleaning of natural stone

Natural stone

Cleaning by water coupled with scrubbing or the use of a high pressure water jet could effectively remove most stains from stone cladding surfaces. Cleaning should begin at the top so that excess water can run down and pre-soften the dirt below. Acidic cleaning agents should not be used for granite as they may attack the pyrite (iron sulphide) which is inherent in granite, and result in brown stains. It is also not proper to use cleaners that contain petroleum (which may change the appearance of the stone) or products that contain other acids or abrasives that may scratch the surface.

Cleaning of aluminium cladding

Tiles

Depending on the tiles used and the extent of staining on the surface, cleaning agents can be selected according to the state of staining. Care should be taken during the selection of cleaning agents, as abrasive agents can easily etch the tiles, making them more vulnerable to dirt.

(*Continued*)

(*Continued*)

Wiping glass surface with a chamois

Cleaning of glass façade surface

Metal

There are many ways to clean metals, from using plain water to harsh abrasives. The mildest possible method should be used first, particularly for anodised aluminium. The following cleaning materials and procedure are categorised according to their level of harshness such as 1) plain water; 2) mild soap or detergent; 3) solvent cleaner (e.g. kerosene, turpentine, white spirit); 3) non-etching chemical cleaner; 4) wax-based polish cleaner; 5) abrasive wax and 6) abrasive cleaner. The mildest treatment should first be tried on a small area and if the results are not satisfactory, the next least harsh method may be examined.

Glass

The cleaning of glazed surfaces begins with wringing a cloth, sponge, or chamois until it is almost dry before wiping the glass surface. The wet surface is then dried with newspapers, paper towels, window wipes, or a chamois. Avoid washing windows in direct sunlight because they tend to streak and are more difficult to clean.

Plaster and paint surfaces

Stained plaster and paint walls are usually cleaned by washing and scrubbing. However, if the stains are too serious and widespread, it is more appropriate to remove the affected surface coating; sand, clean and redo the coating.

3.3 Services

Service components refer to the accessibility, fixtures and fittings of the façade. Each issue is presented with its corresponding design, construction and maintenance measures to ensure better performance.

Problem	Design	Construction	Maintenance
Facade access  Track system	Ease of access, relative cost of hiring and erecting scaffolding and the probable frequency of maintenance should be considered when making decisions on façade work (BS 8221-2:2000, SS 509-2:2005 (2015)). The façade access system should be designed	Implement comprehensive safety plan for working on façade, which must include a fall prevention plan, permit-to-work system and fall control measures (including fall prevention systems and personal fall arrest systems) [14].	The least hazardous product and system should be selected for the façade cleaning operation. All risks should be identified, assessed and managed (BS 8221-1:2012, SS 509-1:2015).

(*Continued*)

(*Continued*)

Problem	Design	Construction	Maintenance
 Maintenance crew using rope access  Façade cleaning automated machine (IFF)  Downhook access system  Floor slab mounted gondola system	according to the following minimum requirements [12]: — should have a long design life of between 10–15 years; — should be able to withstand all loadings (wind, dead and live) as stipulated by local codes and specifications; — should provide easy and safe access to all façade areas for cleaning, repair and replacement works; — should provide maximum coverage to the façade, including clearing protrusions on façades (e.g. sunshades) to fulfill required cleaning cycle.	Provisions of the Workplace Safety and Health Act should be abided on the aspect of safe design, construction and maintenance of scaffolding, working platforms and gondolas (BS 6150:2006+ A1:2014, SS 542:2008). Use scaffoldings and associated components in accordance with BS EN 1004:2004, BS EN 39:2001, BS EN 74-1:2005, BS 1139-2.2:2009+A1:2015, BS EN 12810-2:2003, BS EN 12811-1:2003, SS CP 14: 1996. If permanently installed suspended access equipment should be used, refer to BS 6037-1:2003, BS 5974:2010, SS 598:2014.	The safe use of permanently installed building maintenance units for façade maintenance should be done according to ASME A120.1-2014. Cleaning robots can be used to overcome the dangerous and time consuming nature of cleaning work [12]. Façade access systems should be maintained in strict compliance with the relevant codes and standards (SS 598:2014, BS 6037-1:2003). Façade access systems should not disrupt tenant's activities nor cause damage to the façade when in operation [12].
Fixtures and fittings  Maintenance crew fixing the LED media wall lighting	Comply with accessibility requirements, to allow maintenance personnel to reach fixtures and fittings on the façade.	Serviceability of façade LED lightings. All components' life spans are indicated in the schematics, including their points of failure (e.g. Faulty Façade LED lighting).	Defective downpipes, gutters, flashing, lead coverings, and jointing should be repaired quickly, and obsolete cables and fixings should be removed (BS 8221-2:2000, SS 509-2:2005(2015)).

References

[1] Chew, M. Y. L. (2016). *Maintainability of Facilities: Green FM for Building Professionals* (2nd ed.). Singapore: World Scientific.

[2] NUS Maintainability of Buildings (2017). Façade. Retrieved on March 9 from http://www.hpbc.bdg.nus.edu.sg/?page_id=46.

[3] ASTM International (2014). ASTM E2270-14, Standard Practice for Periodic Inspection of Building Façades for Unsafe Conditions. West Conshohocken, PA: ASTM International.

[4] Hong Kong Government Buildings Department (2017). Mandatory Building Inspection Scheme and Mandatory Window Inspection Scheme — Buildings (Amendment) Bill 2010. Retrieved on May 9 from http://www.bd.gov.hk/english/services/index_MBIS_MWIS.html

[5] Venning, C. (1995). External access systems. In C. Briffet (Ed.), *Building Maintenance Technology in Tropical Climates* (pp. 139–157). Singapore: Singapore University Press.

[6] P. Perkins (2002). *Repair, Protection and Waterproofing of Concrete Structures* (3rd ed.). Florida, USA: CRC Press.

[7] Centre for Window and Cladding Technology (2001). *CWCT Curtain Wall Installation Handbook*. United Kingdom: Centre for Window and Cladding Technology, University of Bath.

[8] Building and Construction Authority (2004). Good Industry Practices — Waterproofing for External Wall. Singapore: Building and Construction Authority (BCA).

[9] Building and Construction Authority (2017). *CONQUAS® The BCA Construction Quality Assessment System* (9th ed.). Singapore: Building and Construction Authority (BCA).

[10] NZMRM (2012). *NZ Metal Roof and Wall Cladding Code of Practice* (2nd ed.). New Zealand: New Zealand Metal Roofing Manufacturers Association Inc (NZMRM).

[11] Building and Construction Authority (2003). *Ceramic Tiling* (2nd ed.). Singapore: BCA.

[12] Building and Construction Authority (2016). Regulation on Daylight Reflectance of Materials used on Exterior of Buildings. Retrieved on May 23, 2017 from https: //www.corenet.gov.sg/media/2013555/circular-on-regulation-on-daylight-reflectance-of-materials-used-on-exterior-of-buildings.pdf.

[13] Chew, M. Y. L. and Tan, P. P. (2003). *Staining of Façades*. Singapore: World Scientific.

[14] Workplace Safety and Health Council, Code of Practice for Working Safely at Heights. Singapore: WSHC, 2013.

Normative References/Standards Referred to for Façade

- AS 3958.1-2007/AMDT 1-2010 - Ceramic tiles Guide to installation of ceramic tiles
- ASME A120.1-2014 — Safety Requirements for Powered Platforms and Traveling Ladders and Gantries for Building Maintenance

- ASTM A380/A380M-13 — Standard Practice for Cleaning, Descaling, and Passivation of Stainless Steel Parts, Equipment, and Systems
- ASTM C1193-16 — Standard Guide for Use of Joint Sealants
- ASTM C1260-14 — Standard Guide for Structural Sealant Glazing
- ASTM C1293-08b(2015) — Standard Test Method for Determination of Length Change of Concrete Due to Alkali-Silica Reaction
- ASTM C1401-14 — Standard Test Method for Determining the Potential Alkali-Silica Reactivity of Combinations of Cementitious Materials and Aggregate (Accelerated Mortar-Bar Method)
- ASTM C1487-02(2012) — Standard Guide for Remedying Structural Silicone Glazing
- ASTM C1496-11 — Standard Guide for Assessment and Maintenance of Exterior Dimension Stone Masonry Walls and Façades
- ASTM C1521-13 — Standard Practice for Evaluating Adhesion of Installed Weatherproofing Sealant Joints
- ASTM C1567-13 — Standard Test Method for Determining the Potential Alkali-Silica Reactivity of Combinations of Cementitious Materials and Aggregate (Accelerated Mortar-Bar Method)
- ASTM C1722-11 — Standard Guide for Repair and Restoration of Dimension Stone
- ASTM C227-10 — Standard Test Method for Potential Alkali Reactivity of Cement-Aggregate Combinations (Mortar-Bar Method)
- ASTM C295/C295M-12 — Standard Guide for Petrographic Examination of Aggregates for Concrete
- ASTM C618-15 — Standard Specification for Coal Fly Ash and Raw or Calcined Natural Pozzolan for Use in Concrete
- ASTM C639-15 — Standard Test Method for Rheological (Flow) Properties of Elastomeric Sealants
- ASTM C856-17 — Standard Practice for Petrographic Examination of Hardened Concrete
- ASTM C920-14a — Standard Specification for Elastomeric Joint Sealants
- ASTM C926-17 — Standard Specification for Application of Portland Cement-Based Plaster
- ASTM C989/C989M-16e1 — Standard Specification for Slag Cement for Use in Concrete and Mortars
- ASTM E903 — Standard Test Method for Solar Absorptance, Reflectance, and Transmittance of Materials Using Integrating Spheres or equivalent.
- ASTM E1300-16 — Standard Practice for Determining Load Resistance of Glass in Buildings
- ASTM E1667-95a(2012)) — Standard Classification for Serviceability of an Office Facility for Image to the Public and Occupants
- ASTM E2112-07(2016) — Standard Practice for Installation of Exterior Windows, Doors and Skylights
- ASTM E2128-12 — Standard Guide for Evaluating Water Leakage of Building Walls

- ASTM E2266-11 — Standard Guide for Design and Construction of Low-Rise Frame Building Wall Systems to Resist Water Intrusion
- ASTM E2270-14 — Standard Practice for Periodic Inspection of Building Façades for Unsafe Conditions
- ASTM E241-09(2014)e1 — Standard Guide for Limiting Water-Induced Damage to Buildings
- ASTM E2513-07(2012) — Standard Specification for Multi-Story Building External Evacuation Platform Rescue Systems
- ASTM E2841-11 — Standard Guide for Conducting Inspections of Building Façades for Unsafe Conditions are now added in the references
- BS 1139-2.2:2009+A1:2015 — Metal scaffolding. Couplers and fittings. Couplers and fittings outside the scope of BS EN 74. Requirements and test methods
- BS 4873:2016 — Aluminium alloy windows and doorsets. Specification
- BS 5427:2016 — Code of practice for the use of profiled sheet for roof and wall cladding on buildings
- BS 5974:2010 — Code of practice for the planning, design, setting up and use of temporary suspended access equipment
- BS 6037-1:2003 — Code of practice for the planning, design, installation and use of permanently installed access equipment. Suspended access equipment
- BS 6093:2006+A1:2013 — Design of joints and jointing in building construction. Guide
- BS 6150:2006+A1:2014 — Painting of buildings. Code of practice
- BS 6213:2000+A1:2010 — Selection of construction sealants. Guide
- BS 6576:2005+A1:2012 — Code of practice for diagnosis of rising damp in walls of buildings and installation of chemical damp-proof courses
- BS 7543 — Guide to durability of buildings and building elements, products and components
- BS 8000-0:2014 — Workmanship on construction sites. Introduction and general principles
- BS 812-123:1999 — Testing aggregates. Method for determination of alkali-silica reactivity. Concrete prism method
- BS 1161:1977 — Specification for aluminium alloy sections for structural purposes
- BS 8215:1991 — Code of practice for design and installation of damp-proof courses in masonry construction
- BS 8213-4:2016 — Windows and doors. Code of practice for the survey and installation of windows and external doorsets
- BS 8221-1:2012 — Code of practice for cleaning and surface repair of buildings. Cleaning of natural stone, brick, terracotta and concrete
- BS 8221-2:2000 — Code of practice for cleaning and surface repair of buildings. Surface repair of natural stones, brick and terracotta
- BS 8298-1:2010 — Code of practice for the design and installation of natural stone cladding and lining. General

- BS EN 1004:2004 — Mobile access and working towers made of prefabricated elements. Materials, dimensions, design loads, safety and performance requirements
- BS EN 1026:2016 — Windows and doors. Air permeability. Test method
- BS EN 1027:2016 — Windows and doors. Water tightness. Test method
- BS EN 1090-2:2008+A1:2011 — Execution of steel structures and aluminium structures. Technical requirements for steel structures
- BS EN 1991-1-1:2002 — Eurocode 1. Actions on structures. General actions. Densities, self-weight, imposed loads for buildings
- BS EN 1991-1-4:2005+A1:2010 Eurocode 1 — Actions on structures. General actions. Wind actions
- BS EN 1992-1-1:2004+A1:2014 — Eurocode 2: Design of concrete structures. General rules and rules for buildings
- BS EN 1993-1-1:2005+A1:2014 Eurocode 3 — Design of steel structures. General rules and rules for buildings
- BS EN 1999-1-1:2007+A2:2013 Eurocode 9 — Design of aluminium structures. General structural rules
- BS EN 1999-1-4:2007+A1:2011 Eurocode 9 — Design of aluminium structures. Cold-formed structural sheeting
- BS 8213-4:2016 — Windows and doors. Code of practice for the survey and installation of windows and external doorsets
- BS EN 10088-2:2014 — Stainless steels. Technical delivery conditions for sheet/plate and strip of corrosion resisting steels for general purposes
- BS EN 10346:2015 — Continuously hot-dip coated steel flat products for cold forming. Technical delivery conditions
- BS EN 12152:2002 — Curtain walling. Air permeability. Performance requirements and classification
- BS EN 12153:2000 — Curtain walling. Air permeability. Test method
- BS EN 12154:2000 — Curtain walling. Water-tightness. Performance requirements and classification
- BS EN 12179:2000 — Curtain walling. Resistance to wind load. Test method
- BS EN 12810-2:2003 — Façade scaffolds made of prefabricated components. Particular methods of structural design
- BS EN 12811-1:2003 — Temporary works equipment. Scaffolds. Performance requirements and general design
- BS EN 13022-1:2014 — Glass in building. Structural sealant glazing. Glass products for structural sealant glazing systems for supported and unsupported monolithic and multiple glazing
- BS EN 13369:2013 — Common rules for precast concrete products
- BS EN 13561:2015 — External blinds and awnings. Performance requirements including safety
- BS EN 13830:2015 — Curtain walling. Product standard
- BS EN 13914-1:2016 — Design, preparation and application of external rendering and internal plastering. External rendering
- BS EN 14992:2007+A1:2012 — Precast concrete products. Wall elements

- BS EN 15651-1:2017 — Sealants for non-structural use in joints in buildings and pedestrian walkways. Sealants for façade elements
- BS EN 1932:2013 — External blinds and shutters. Resistance to wind loads. Method of testing and performance criteria
- BS EN 1992-1-1:2004+A1:2014 — Eurocode 2: Design of concrete structures. General rules and rules for buildings
- BS EN 1993-1-4:2006+A1:2015 — Eurocode 3. Design of steel structures. General rules. Supplementary rules for stainless steels
- BS EN 1994-1-1:2004 — Eurocode 4. Design of composite steel and concrete structures. General rules and rules for buildings
- BS EN 1995-1-1:2004+A2:2014 — Eurocode 5: Design of timber structures. General. Common rules and rules for buildings
- BS EN 39:2001 — Loose steel tubes for tube and coupler scaffolds. Technical delivery conditions
- BS EN 485-1:2016 — Aluminium and aluminium alloys. Sheet, strip and plate. Technical conditions for inspection and delivery
- BS EN 74-1:2005 — Couplers, spigot pins and baseplates for use in falsework and scaffolds. Couplers for tubes. Requirements and test procedures
- BS EN ISO 12631:2012 — Thermal performance of curtain walling. Calculation of thermal transmittance
- BS EN ISO 12944-8:1998 — Paints and varnishes. Corrosion protection of steel structures by protective paint systems. Development of specifications for new work and maintenance
- ISO 13823:2008 — General principles on the design of structures for durability
- ISO 15099:2003 — Thermal performance of windows, doors and shading devices — Detailed calculations
- ISO 15686-1:2011 — Buildings and constructed assets — Service life planning -- Part 1: General principles and framework
- ISO 28278-2:2010 — Glass in building — Glass products for structural sealant glazing — Part 2: Assembly rules
- ISO 28841:2013 — Guidelines for simplified seismic assessment and rehabilitation of concrete buildings
- ISO 6589:1983 — Joints in building — Laboratory method of test for air permeability of joints
- ISO 7361:1986 — Performance standards in building — Presentation of performance levels of façades made of same-source components
- ISO 7729:1985 — Typical vertical joints between two prefabricated ordinary concrete external wall components — Properties, characteristics and classification criteria
- ISO 7845:1985 — Horizontal joints between load-bearing walls and concrete floors — Laboratory mechanical tests — Effect of vertical loading and of moments transmitted by the floors
- ISO/NP 16938-1 — Buildings and civil engineering works — Determination of the staining of porous substrates by sealants used in joints — Part 1: Test with compression
- SS 212:2007 — Specification for aluminium alloy windows

- SS 509-1:2015 — Code of practice for cleaning and surface repair of buildings — Part 1: Cleaning of natural stone, brick, terracotta, concrete and rendered finishes
- SS 509-2:2005(2015) — Code of practice for cleaning and surface repair of buildings — Surface repair of natural stones, brick, terracotta and rendered finishes
- SS 542:2008 — Code of practice for painting of buildings
- SS 555 : Parts 1 to 4 : 2010 — Code of practice for protection against lightning
- SS 598:2014 — Code of practice for suspended scaffolds
- SS 599:2014 — Guide for wayfinding signage in public areas
- SS CP 14:1996 — Code of practice for scaffolds
- SS CP 65-1:1999 — Code of practice for structural use of concrete — Design and construction
- SS CP 65-2:1996(1999) — Code of practice for structural use of concrete — Special circumstances
- SS CP 68 : 1997 - Code of practice for ceramic wall and floor tiling
- SS CP 7:1997(2014) — Code of practice for structural use of timber
- SS CP 81:1999 — Code of practice for precast concrete slab and wall panels
- SS CP 82:1999 — Code of practice for waterproofing of reinforced concrete buildings
- SS CP 96:2002(2011) — Code of practice for curtain walls
- SS EN 1992 — 1-1:2008 — Eurocode 2: Design of concrete structures, Part 1-1 General rules and rules for buildings
- SS EN 1992 — 1-2:2008 (2015) — Singapore National Annex to Eurocode 2: Design of concrete structures — Part 1-2 General rules — Structural fire design
- SS EN 1994-1-2:2009 — Eurocode 4 — Design of composite steel and concrete structures — General rules — Structural fire design

Chapter

Roofs

Introduction

Roofs function as a shelter over one's head. It serves to enclose a space, prevent the penetration of inclement weather, and control heat gain or heat loss. They may be classified as flat roofs, pitched roofs and curved roofs.

Flat roofs (roofs that slope at less than 10 degrees) are usually selected for storage, recreation and maintenance uses, while pitched roofs are chosen for their easy design and construction. A pitched roof's slope allows it to shed water quickly, while the airspace within the roof envelope provides good thermal insulation. Curved roofs (e.g. shell roof, vaulted roof, dome, etc.) are selected primarily for artistic and architectural expression. Shell roofs are three-dimensional structures built with an arched solid slab or membrane functioning as a stressed skin to transmit loading to a point of support [1].

Roofs should have a minimum fall of 1:40 to drain out the water. The prevalent defects of flat roofs are water leakages due to drainage problems and the wear and tear of roofing materials (e.g. waterproofing membrane). Discontinuity is the main weak point of any waterproofing system; where the membrane is either interrupted, terminated, or at joining of an area or projection, has a marked change in slope or direction.

Another important concern is the occurrence of cracking on roofs, which can lead to water leakage and accommodate the easy penetration of destructive elements to the steel reinforcement's level, resulting in corrosion. Common roof defects are identified as: (a) water ponding and leakage, (b) sealant failure, (c) blistering, (d) premature membrane failure, (e) cracking. Additional roof defects are leakage

View of a shell roof from the garden area (Esplanade, Singapore).

through pipe penetrations and tile bursting, while water ponding on a roof may result in water leakage and algae and fungal growth due to the deterioration of the waterproofing system [2].

Relevant guidelines for the design, construction and maintenance of roofs are highlighted in the succeeding tables, which provide detailed specifications to aid in understanding the prevalent defects and prevent them from occurring during a roof's entire lifespan. The structural defects refer to cracks at drainage and expansion joints, and leakage through slabs. The architectural defects centre around water ponding and waterproofing issues. The service defects pertain to roof access, penetration for services, drainage, and issues with fixtures and fittings.

Roof durability is attributed to material selection and workmanship, while the main concern in the design and construction of durable roofs lies in the roof's compatibility and performance within the system. Upturns, flashings, overlaps, etc. are proven methods to counteract such waterproofing discontinuity. Sealant should be provided around all pipes to help prevent water from seeping in during any differential movement among elements. The drainage system should be designed as per local standard and weather data, as well as fulfil two basic requirements: (a) consideration for the strength and construction of the roof; and (b) the operating head of an outlet should not cause a build-up of water that exceeds the roof's designed loading nor penetrate the roof covering [3].

Correct roof construction practices may include: (a) surface preparation; (b) proper lapping and ensuring that preformed waterproofing membrane is applied with a minimum of 75mm end laps and 50mm side laps; and (c) the proper application of either preformed membrane (torch-on-method) or liquid applied membrane. The proper maintenance and usage of a roof can enhance its life significantly and delay the almost inevitable process of re-roofing to a certain extent. A typical roof houses various mechanical equipment (e.g. the base of the cleaning gondola, cooling tower, overhead water storage tanks, etc.) and may deteriorate due to the substandard design, construction or maintenance of such ancillary building services. Heavy equipment may cause local depressions, or rust dripping from corroded parts, as well as ducts penetrating through the roof deck may contribute to the discontinuity in waterproofing. Providing roof accessibility and ensuring safety should also be considered at the design, construction and maintenance stages to avoid falls and other physical damages and injuries [4].

4.1 Structural

The structural components of a roof deck require adequate provisions in terms of design, construction and maintenance to prevent premature deterioration and to achieve optimum performance under specific external environmental exposure.

Problem	Design	Construction	Maintenance
Roof leakage (through cracks at drainage/expansion joints and through slabs) 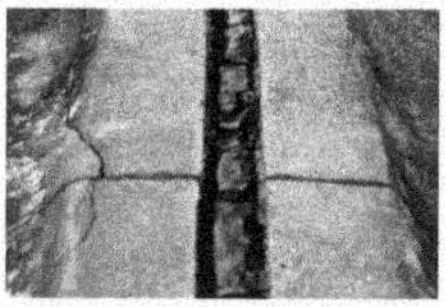 Crack at expansion joint  Sealant failure (bulging)	To prevent potential construction and maintenance issues, consider good design practices such as allowing proper flow paths to suitable discharge points, specifying suitable waterproofing membrane, and using the necessary movement joints. Properly detail all openings, penetrations, upturns, and corners to ensure water-tightness (BS EN 1107-2: 2001, BS EN 12039:2016, BS EN	Use low permeability concrete for roof structure construction. Mix concrete appropriately (maintain specified water–cement ratio) and ensure proper curing to prevent excessive evaporation, which can lead to defects. Consider storing aggregates under shade (SS EN 1992-1-1:2008, SS EN 1992-1-2:2008 (2015), SS CP 65-1:1999). Waterproofing and drainage must be coordinated to form an	Conduct inspection to determine the current roof condition and document visible evidence of water leakage. Indications of wear and tear, maintenance, attempted repairs, damage from non-weather related causes such as impacts, or structural movements must be recorded (ASTM D7053/ D7053M-17). Sealants can crack/lose adhesion at its interface under solar radiation (ISO 13823:2008).

(*Continued*)

(*Continued*)

Problem	Design	Construction	Maintenance
 Crack perpendicular to the flow of the scupper drain 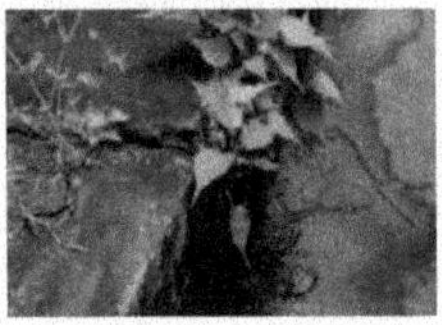 Plant growth on crack at sealant joint of roof drain 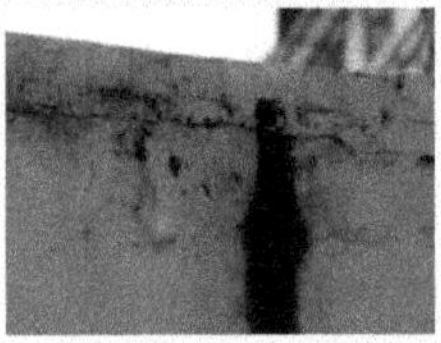 Cracks, paint peeling and water leak along roof slab gutter 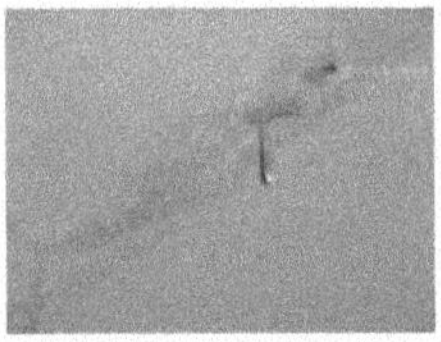 Efflorescence formation, paint peeling and possible fungal growth	12730:2015, SS CP 82:1999). Provide waterproofing (materials and details) to prevent water leakage through external joints (BS EN 15037-5:2013, BS EN 13693:2004+ A1:2009, BS EN 12730:2015, BS EN 13369:2013, SS CP 81: 1999). To avoid the risk of movement being transferred to the waterproofing membrane, detail expansion joints as per BS 093:2006+ A1:2013. Refer to the code of practice for structural use of concrete (SS EN 1992-1-1:2008, SS EN 1992-1-2:2008 (2015), SS CP 65-1:1999) for RC roof design. For precast roof slab design, refer to the guidelines as per BS EN 15037-5:2013, BS EN 13693:2004+ A1:2009, BS EN 13369:2013, SS CP 81:1999.	integrated waterproofing system. Maintain construction quality control to ensure proper curing of concrete prior to membrane application. Ensure that reinforced concrete roof and waterproofing membrane do not trap excessive moisture during construction. Quality of workmanship on roof construction to comply with BS 8000-0:2014. Consistency of construction quality in terms of; installation of fixtures and fittings; proper dressing of pipe penetrations; and good lapping and adhesion of waterproofing membrane to base with no mortar stains. Check for any signs of leakage during construction [5]. The sealant material used must be tolerant of climatic variations on-site and be able to accommodate high movement if necessary. Joint sealing slots need to be cleaned	Check for roof leaks yearly, especially at roof penetrations (ASTM E241-09(2014)e1). Use thermography to identify position of water leakages on roof. Gutters and downspouts must be maintained to enable free/undisrupted flow of water and be kept clear of leaves and other debris to avoid occurrences of excessive roof runoff and eventual leakage. Review work orders and purchase orders for building maintenance and other activities that may relate to water leakage problems. Maintain service history: the known performance record of the roof system, including the physical symptoms of water leakage, progression of leakage behaviour, maintenance and repair history, extent and locations of leakage, etc. (ASTM D7053/ D7053M-17).

(*Continued*)

(*Continued*)

Problem	Design	Construction	Maintenance
 Roof gutter leakage	Ensure positive drainage in all low-slope roofs [6]. Design roof gutter according to the performance requirements set by BS EN 12056-3:2000, SS 525:2006. Roofing components should follow the allowable thermal stresses under temperature variations (ISO 13823:2008).	(i.e. remove all dust). Proper installation of backer rod necessary to avoid sealant adhering to the base of the slot. Tightly pack joint filler to slot. Consider providing an upturn at expansion joints (with metal capping). Expansion joints should be properly constructed as per BS 6093:2006+A1:2013.	Repair materials/ methods should be designed to manage environmental factors (e.g. chemical/ physical/ mechanical conditions). Adhere to the specifications of grouting materials for repair as set under ISO/TR 16475:2011.

4.2 Architectural

Architectural components of a roof deck (e.g. floor finishes and waterproofing system) contribute to its aesthetic value and durability.

Problem	Design	Construction	Maintenance
Water ponding  Lack of adequate drainage due to lack of roof gradient/fall	Flat roofs made of reinforced concrete should be designed with an adequate fall to prevent ponding and maintain minimum pitch variation over the entire roof to avoid water ponding at the perimeter and water leakage through the roof slab/parapet wall junction (BS 6229:2003, BS 8218:1998, SS CP 82:1999).	Ensure quality waterproofing by maintaining adequate and correct fall directions and angles to gutters (BS EN 12056-3:2000, BS 8490:2007, BS 8000-0:2014). The waterproofing and drainage must be coordinated to form an integrated waterproofing system. Conduct visual inspections and ensure that all gutters are	Flat roofs should be inspected for ponding. Visual observations of free flowing water towards the outlets must be made to ensure that the water drains off thoroughly, so as to avoid leftover ponding in the gutter or on the reinforced concrete flat roof (BS 8221-1:2012, SS 509-1:2015). Free flow in gutters and downspouts must be maintained. They must

(*Continued*)

(*Continued*)

Problem	Design	Construction	Maintenance
 Rooftop water ponding	Ensure positive drainage in all low-slope roofs. Do not obstruct joints of areas prone to water, ice and freeze/thaw damage [6].	tested for ponding upon construction (BS EN 12056-3:2000, BS 8000-0:2014, SS 525:2006).	be kept clear of leaves and other debris to avoid exposure to excessive roof runoff and eventual leakage.
Waterproofing 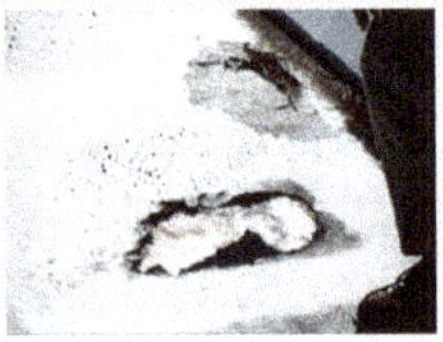 Bursting of blistered membrane  Alligatoring  Delamination	Conform to design considerations to prevent potential construction and maintenance issues.[1] Select proper waterproofing membrane for roof, with consideration for type of roof loading, environmental exposures, thermal insulation and aesthetics. Thickness of membrane as per manufacturer recommendation (normally > 1.5mm). Provide proper detailing for bridging of membrane over expansion joints (BS EN 1107-2:2001, BS EN 12039:2016, BS EN 12730:2015, BS EN 1108:2000, SS CP 82:1999).	Before laying the liquid applied membrane, ensure that the surface is cleaned, dried, and free of surface defects/sharp protrusions (concrete surface to have a plain and even finish prior to receiving the membrane). Ensure that all openings, penetrations, upturns, corners, etc. are properly installed to ensure water-tightness. Additional consideration should be given to flashing and kerbs for movement. Allow concrete/screeding to cure for at least 7 days prior to laying of the membrane.[2] Waterproofing membrane should be	Good quality waterproofing membrane requires minimal maintenance. However, it depends mainly on its exposure conditions and external factors. Conduct regular inspections and cleaning of roof surface. Roof area drains should be inspected for efficiency prior to water washing (BS 8221-1:2012, SS 509-1:2015). Diagnosis of roof membrane defects can be done using infra-red thermographic surveys which identify moisture build-up within the structure. Diagnosis of waterproofing

(*Continued*)

[1]See *Maintainability of Buildings, Rooftop. Available at: http://www.hpbc.bdg.nus.edu.sg/?page_id=64 (accessed May 12, 2017)* for detailed information on different roof construction types (e.g. concrete deck, metal, pitch) and their respective design and construction considerations.

[2]See *Maintainability of Buildings, Blistering of Waterproofing Membrane. Available at: http://www.hpbc.bdg.nus.edu.sg/?page_id=894 (accessed May 12, 2017)* for good practices on how to prevent blistering from occurring.

(*Continued*)

Problem	Design	Construction	Maintenance
 Delaminated waterproofing membrane from joints	Conduct evaluation of preformed waterproofing membranes for roofs. Refer to proper specifications in waterproofing (BS 8217:2005, BS 8747:2007, SS 374:1994, SS 133:1987 (1998)).	protected from traffic and weathering. It should be applied soon after the primer is cured (BS 6229:2003, BS 8218:1998, BS 8000-0:2014, SS CP 82:1999).	membrane failures[3] leads to remedial actions if required. Actions can range from patch repairs to re-roofing. It can also include creating effective slope, or adding roof drains or taper systems.

4.3 Services

Service components, here, refer to the accessibility and plumbing–sanitary connections of a roof. Each issue is presented with its corresponding design, construction and maintenance measures to ensure better performance.

Problem	Design	Construction	Maintenance
Roof access  Maintenance crew using rope access	Provide safe access to building roof, and if required, fall protection, and building maintenance units during design.	Identify fall hazards at different stages of work on roofs [4]. Use personal fall arrest systems during construction (BS EN 363:2008, BS EN 354:2010, BS EN 361:2002, SS 528 Series, SS 570:2011).	Develop and implement safe maintenance work plan for any activities at roof areas. Proper supervision of workers at heights is recommended. Provide maintenance workers with safe means of getting to and from the roof. The accessways

(*Continued*)

[3]See *Maintainability of Buildings, Premature Membrane Failure. Available at: http://www.hpbc.bdg.nus.edu.sg/?page_id=895 (accessed May 12, 2017)* for more on waterproofing membrane defects viz. alligatoring, delamination, adhesive failure, tearing, splitting, and ridging, and how to diagnose and remedy those defects.

(*Continued*)

Problem	Design	Construction	Maintenance
 Maintenance worker cleaning the rooftop garden  Roof anchor for maintenance works	Conduct risk assessment and designate controlled access zones to design for proper safety systems. Provide anchorage for personal fall arrest systems where necessary (e.g. permanent horizontal anchor-cable systems). Consider the usage of roof access equipment for fragile roofs (e.g. board walk, roof walk systems).	Use scaffoldings as per SS CP 14:1996. Properly construct and install safety systems. Provide demarcation of access routes plan for equipment maintenance. For new roof installations or where extensive repair or replacement of existing roof is planned, it is recommended that an access tower or a personnel and materials hoist be provided [7].	need to be in place before commencing work. Access should be located where the work on the roof is to begin [4]. Use safety systems during roof repair work, including the roof anchorage provided for PPE (personal protective equipment) to be attached to, and travel restraint systems with other fall prevention methods such as guard-rails and personal fall arrest systems for roof works [7]. Train maintenance personnel on proper usage of fall protection systems (BS EN 363:2008).
Penetration for services  Badly designed pipe protrusion  Pipe protrusion along wall and floor joints leading to water seepage under slab	Joints at pipe penetrations/fittings/ precast elements, etc., should be designed with shapes and dimensions that do not constitute local weak points in the elements (e.g. at corners). Avoid joints that are difficult to construct. Penetrations should be kept at a minimum as far as possible. It is preferable to have a clear and uninterrupted roof deck for continuous waterproofing. Services should be grouped, pre-planned and boxed out	Waterproofing membrane at pipe penetrations must be applied with an upturn onto the pipes. In precast construction, a cast in sleeve is used to avoid pipe openings which have to be cast back after pipe installation (BS 6229:2003, BS 8218:1998, BS 8000-0: 2014, SS CP 82:1999). Once installed, ensure that the membrane is protected from damage by construction activities and detrimental exposure conditions.	Periodic walkthrough to check roof and underside of roof deck areas (special care needs to be made at hard to access areas) to identify signs of water seepage. A thermo-tracer can be used to locate moisture accumulating at pipes penetrations. Check for roof leaks yearly, especially at roof penetrations (ASTM E241-09(2014)e1). Adhere to waterproofing guidelines for upstands and penetration during roof maintenance and

(*Continued*)

(*Continued*)

Problem	Design	Construction	Maintenance
 Penetrations at roof seals	to minimise penetration through waterproofing (BS 6229:2003, BS 8218:1998, SS CP 82:1999). Penetration of services in any roof system are most susceptible to water entry; thus the design of all flashing details[6].	A protection system that is quick and easy to install is recommended. Pipe penetrations should be effectively fire-stopped by replacing the insulation material at the junction (BS EN 13501-5:2016, SS 553:2016).	repair work (BS 6229:2003, BS 8218: 1998, BS 8000-0:2014, SS CP 82:1999).
Drainage  Blockage at drain outlet fixture and a snail blocking the drain hole  Poorly maintained roof drain  Maintenance workers repairing roof drainage	Calculate the effective catchment area for the roof and rate of run-off from the roof to decide the proper materials and components for rainwater goods (BS 460:2002+A2:2007, BS EN 12056-3:2000, SS 525:2006). Provide required number of internal rainwater outlets and drainage downpipes. Specify for minimum falls of 1:40 for flat roofs to ensure a 1:80 finished fall (BS 6229:2003, BS 8217:2005). Conform to drainage requirements of the building (floor gullies and roof drains) as per BS EN 1253-2:2015, BS EN 476:2011. Comply with manufacturer's details during design for roof components such as drains and base flashings to enhance roof's buildability. Ensure positive drainage in all low-slope roofs. Do not obstruct joints of areas prone to water damage [6].	Ensure that rainwater pipes and gutters have adequate capacity (BS 6229:2003, BS EN 12056-3:2000, BS 8490:2007, BS 8000-0: 2014, SS CP 82:1999). Roof covering should not extend too far into gutter, so as to accommodate easy cleaning and maintenance. Jointing type for gutters and rainwater pipes should allow thermal movement to take place without leakage, or distortion and displacement of fittings (BS EN 12056-3:2000, SS 525:2006). Service fixtures, planters etc. may be designed to stand over the waterproofed deck on concrete pads (BS 6229:2003, BS 8218:1998, BS 8000-0:2014, SS CP 82: 1999).	Rainwater pipes, gutters, and outlet gratings should be inspected and thoroughly cleaned at least once a year or more, especially if the building is near an industrial area or trees, or subjected to extreme temperature differences (BS EN 12056-3:2000, SS 525:2006). Clear gutters and downspouts, if and when needed. Frequency depends on proximity of trees to the building. Special attention should begiven to birds' nests, as they may cause blockages (ASTM E241-09(2014)e1). Adhere to proper maintenance to prevent mosquito breeding. Prevent water stagnation in roof drainage systems and avoid ponding. Maintain safe and permanent roof access for maintenance purposes (BS EN 12056-3:2000, SS 525:2006).

SOME EXAMPLES OF ROOF DEFECTS AND REPAIRS [2]

Splitting

Gravel Scouring

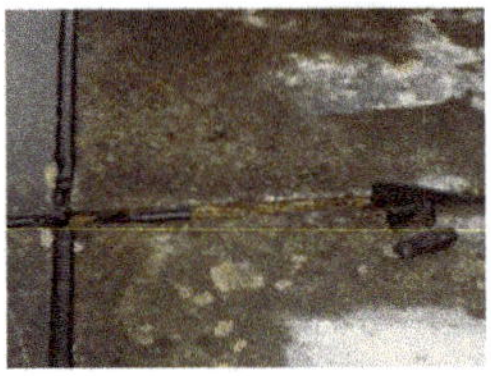

Blistering

Splitting — Tends to occur at points of weakness, such as joints, corners or roof penetrations. Thus making it necessary to replace damaged insulation and/or re-roof the affected portion.

Gravel Scouring — Typically caused by exposure of roofing membrane to extreme weathering such as high winds, or water on protected membrane roofs. The insulation will be exposed to weathering once the gravel cover is lost. This defect can be addressed by replacing and re-roofing the affected portion.

Blistering — Often caused by moisture in the roof system. It occurs in the coating and results in the exposure of felt to weathering. For remedial purposes, check and replace the affected membrane, as well as recoat and re-gravel.

Slippage — Signs of slippage of membranes on conventional sloping roofs are exposed bitumen on higher parts of the roof and piling up of felts on lower ones. Slippage happens as a result of the bitumen having too low a viscosity for the slope and is likely to occur in warm weather when roof temperatures may reach 70°C or more. The viscosity of the bitumen may increase enough with age to stabilize it. Alternatively, it may be possible to secure the membrane to the insulation (which is quite probably stable) or the deck using some mechanism. It can also occur on protected membrane roofs if the slope and top-surface loading combine to produce a sufficiently high shear stress on the membrane, even though membrane temperatures will probably not exceed 35°C. Blocking, secured to the deck, may be required to prevent it. This defect can be addressed by replacing and re-roofing the affected portion [8].

References

[1] Chew, M. Y. L. (2017). *Construction Technology for Tall Buildings* (5th ed.). Singapore: World Scientific.

[2] NUS Maintainability of Buildings (2017). Rooftop. Retrieved on March 9 from http:// www. hpbc.bdg.nus.edu.sg/?page_id=50.

[3] Chew, M. Y. L. (2016). Maintainability of Facilities: Green FM for Building Professionals (2nd ed.). Singapore: World Scientific.

[4] Workplace Safety and Health Council (2013). *Workplace Safety and Health Guidelines: Working safely on roofs*. Singapore: WSHC.

[5] Building and Construction Authority (2017). *CONQUAS® The BCA Construction Quality Assessment System* (9th ed.). Singapore: Building and Construction Authority (BCA).

[6] Watson, D. and Crosbie, M. J. (2005). *Time-Saver Standards for Architectural Design Data: Technical Data for Professional Practice* (8th ed.). USA: McGraw-Hill.

[7] Workplace Safety and Health Council (2013). *Code of Practice for Working Safely at Heights*. Singapore: WSHC.

[8] NUS Maintainability of Buildings (2016). Reinforced Dense Concrete: Defects and Repair. Material Manual. Retrieved on January 22 from www.hpbc.bdg.nus.edu.sg/? page_id= 392& page=5.

Normative References/Standards Referred to for Roof

- ASTM D7053/D7053M-17 — Standard Guide for Determining and Evaluating Causes of Water Leakage of Low-Sloped Roofs
- ASTM E241-09(2014)e1 — Standard Guide for Limiting Water-Induced Damage to Buildings
- BS 460:2002+A2:2007 — Cast iron rainwater goods. Specification
- BS 6093:2006+A1:2013 — Design of joints and jointing in building construction. Guide
- BS 6229:2003 — Flat roofs with continuously supported coverings. Code of practice
- BS 8000-0:2014 — Workmanship on construction sites. Introduction and general principles
- BS 8217:2005 — Reinforced bitumen membranes for roofing. Code of practice
- BS 8218:1998 — Code of practice for mastic asphalt roofing
- BS 8221-1:2012 — Code of practice for cleaning and surface repair of buildings. Cleaning of natural stone, brick, terracotta and concrete
- BS 8490:2007 — Guide to siphonic roof drainage systems
- BS 8747:2007 — Reinforced bitumen membranes (RBMs) for roofing. Guide to selection and specification
- BS EN 1107-2:2001 — Flexible sheets for waterproofing. Determination of dimensional stability. Plastic and rubber sheets for roof waterproofing

- BS EN 1108:2000 — Flexible sheets for waterproofing. Bitumen sheets for roof waterproofing. Determination of form stability under cyclical temperature changes
- BS EN 12039:2016 — Flexible sheets for waterproofing. Bitumen sheets for roof waterproofing. Determination of adhesion of granules
- BS EN 12056-3:2000 — Gravity drainage systems inside buildings. Roof drainage, layout and calculation
- BS EN 1253-2:2015 — Gullies for buildings. Roof drains and floor gullies without trap
- BS EN 12730:2015 — Flexible sheets for waterproofing. Bitumen, plastic and rubber sheets for roof waterproofing — Determination of resistance to static loading
- BS EN 13369:2013 — Common rules for precast concrete products
- BS EN 13501-5:2016 — Fire classification of construction products and building elements. Classification using data from external fire exposure to roofs tests
- BS EN 13693:2004+A1:2009 — Precast concrete products. Special roof elements
- BS EN 15037-5:2013 — Precast concrete products. Beam-and-block floor systems. Lightweight blocks for simple formwork
- BS EN 354:2010 — Personal fall protection equipment. Lanyards
- BS EN 361:2002 — Personal protective equipment against falls from a height. Full body harnesses
- BS EN 363:2008 — Personal fall protection equipment. Personal fall protection systems
- BS EN 476:2011 — General requirements for components used in drains and sewers
- ISO 13823:2008 — General principles on the design of structures for durability
- ISO/TR 16475:2011 — Guidelines for the repair of water-leakage cracks in concrete structures
- SS 133:1987(1998) — Bituminous emulsion for roof waterproofing
- SS 374:1994 — Preformed waterproofing membranes for concealed roof
- SS 509-1:2005 — Code of practice for cleaning and surface repair of buildings — Part 1: Cleaning of natural stone, brick, terracotta, concrete and rendered finishes
- SS 525:2006 — Code of practice for drainage of roofs
- SS 528 Series — Personal fall arrest systems
- SS 553:2016 — Code of practice for air-conditioning and mechanical ventilation in buildings
- SS 570:2011 — Specification for personal protective equipment for protection against falls from a height — Single point anchor devices and flexible horizontal lifeline systems
- SS 598:2014 — Code of practice for suspended scaffolds
- SS CP 14:1996 — Code of practice for scaffolds
- SS CP 65-1:1999 — Code of practice for structural use of concrete — Design and construction
- SS CP 81:1999 — Code of practice for precast concrete slab and wall panels
- SS CP 82:1999 — Code of practice for waterproofing of reinforced concrete buildings
- SS CP 89:2001 — Code of practice for metal roofing
- SS EN 1992-1-1:2008 — Eurocode 2: Design of concrete structures, Part 1-1 General rules and rules for buildings
- SS EN 1992-1-2:2008 (2015) — Eurocode 2: Design of concrete structures, Part 1-2 General rules — Structural fire design

Chapter

Common Areas

Introduction

Common areas, in this chapter, refer to outdoor and indoor open spaces, staircases (including elevators), landscaping and swimming pools.

A common area is a shared space that is available for communal use by all tenants and their visitors, as in the case of condominium developments, as well as for public use in mixed-use commercial development cases. Common areas are set aside for interaction and bonding through recreation, and should be made safe and accessible, and kept well-maintained at all times [1]. A universal design approach, which caters for the broadest range of users from the outset, can result in buildings and places that can be used and enjoyed by everyone; where people of diverse capabilities are able to use these facilities comfortably and safely without special assistance [2]. The design, construction and maintenance of common areas must adhere to relevant legislation and regulations, including those that relate to the care of diverse peoples. This chapter provides guidelines that address each common area defect to prevent them from occurring during the post-occupancy stage. Proper design and material selection, quality workmanship and routine inspections will aid in maintaining a safe, convenient and sound common area.

Outdoor open spaces play an important role for active and passive recreation, and relaxation, and as a playground for children [3]. They provide a setting that simultaneously offers opportunities for social interaction and children's play amongst

Common area of a shopping mall in Singapore.

natural vegetation, as well as help preserve wildlife habitats. Outdoor open space facilities (e.g. accessible pathways, parks and communal areas, bus stops, ramps and stairs) should consider space interconnectivity and building interface. The defects in these areas that have been identified in this chapter, centre around issues with universal design, drop off points, outdoor lighting, and incidents of users slipping and falling. The convenience and safety of drivers and their passengers in accessing a building, as well as the designs of children play areas are critical issues that must be considered in the design, construction and maintenance of a building. This is to ensure that the needs of the public, of all ages and abilities, are catered to, and that a safe environment is provided.

An indoor open space is an interior social space that is generally open and accessible to people of all levels and/or a space constructed for the convenience of all users. Examples of common indoor spaces are lobbies, corridors, stairways, parking lots, ramps, elevators, driveways and storerooms. The defects identified concern issues pertaining to wayfinding/directory and advertisement boards, ventilation, indoor lighting, and universal design. It is important to consider safety precautions and users' differing pace of movement in the design of indoor spaces, to cater to the convenience of all users. A thoughtful building arrangement and layout is crucial for getting people to their destinations and facilitating the smooth flow of human traffic via well-designed routes and signage.

Staircases are important pedestrian routes between different levels. They often also serve as the means of egress from structures and buildings [4]. Because they are constantly subjected to wear and tear due to human and/or goods transportation traffic, it is vital for personnel overseeing the management of facilities to maintain and inspect them, to ensure safety as well as identify any minor or major damage [5]. The identified defects refer to issues concerning a stairway's general requirements, handrails, slippery access routes and accessibility. The design of escape staircases should provide all occupants with a safe and efficient exit. Accessible refuge areas are also vital for occupants to evacuate in any emergency [6]. Additional information regarding the accessibility considerations for elevators are also included — deeper details regarding vertical transportation are dealt with in Chapter 6 of this book.

Landscaping is a task that combines science and art. It refers to the planning, laying out of and construction of gardens that enhance the appearance around a home, institutional setting or estate, while creating a useable space for outdoor activities [7]. The identified defects refer to issues with green roofs/green walls and chokage/discharge issues. The design, construction and maintenance of landscape areas should provide safe, accessible and sound facilities (e.g. paths, pavilions, seats, gardens) to cater the needs of all users.

Swimming pools have become popular venues for relaxation, recreation and social interaction. Their inclusion to a home or estate also adds an appealing feature and increase the value the property [8]. Building and maintaining a pool entails high costs to keep it safe, sanitised and well-kept. Pool water is commonly treated with chlorine to control algae growth and facilitate cleansing [9, 10]. The identified defects pertain to incidents of users slipping and falling, access and egress, building materials, water quality issues and the occurrence of sharp edges due to breakage of tiles. Proper design of and quality workmanship on pools will provide a safer environment for all users. The effective management of health and safety in any swimming pool starts with careful design. Regular and correct maintenance will help eliminate potential hazards [9, 10].

5.1 Outdoor Open Spaces

Problem	Design	Construction	Maintenance
Universal design for outdoor open spaces  Ramp slope in this photo is too steep for wheelchair use  Connecting ramp lacks grab bars/handrails on one side	Conform to the design guidelines for accessibility, ingress/egress to/from building, kerb ramps, accessible routes, horizontal circulation, doors, etc. as per Code on accessibility [11]. (See also BS 8300:2009+A1:2010 and BS 9266:2013). Provide slip-resistant and threshold-free layout, avoid sudden changes in surface levels; design ramps with appropriate slopes and adequate space, elevators and other lifting systems, and stairs with appropriate dimensions and hand grips, to facilitate accessibility (ISO/IEC Guide 71:2014).	Comply to the construction requirements for making buildings accessible to persons with disabilities and families with young children as per Code on Accessibility [11]. (See also BS 8300:2009+A1:2010 and BS 9266:2013). Outdoor paths should be firmly constructed with a slip-resistant, even surface, free from drainage gratings (ISO/NP 21542). Surface material adjacent to the path should not display different slip-resistant characteristics [11]. (See also BS 8300:2009+A1:2010 and BS 9266:2013).	Regular inspection of premises and subsequent correction of irregularities is required for effective housekeeping practices. Clear passageways to allow easy access during firefighting. Outdoor storage areas should be located separately from building to prevent fire spread [12]. Daily/frequent inspection of accessibility and horizontal circulation routes (e.g. look out for broken/detached tiles, blocked and dirty accessibility routes for wheelchairs, etc.) and perform prompt corrective maintenance [11]. (See also BS 8300:2009+A1:2010 and BS 9266:2013).
Poorly-designed drop-off areas  The drop-off point shown here lacks kerb ramp; inhibiting wheelchair accessibility	A designated drop-off zone, for alighting and boarding, should either be provided with a common level or ramp [13]; be sheltered from the weather [1]; and comply with the requirements set by the Code of Accessibility [11]. (See also BS 8300:2009+A1:2010	In the absence of kerb ramps, which provide separation marking between pedestrian and vehicle zones, the construction of a tactile strip (at least 0.60 m wide) at the edge of the pathway is necessary to provide a transition warning to a vehicular area [15, 16].	Keep covered walkways free from obstructions and perform routine mopping/cleaning of water especially during/after rain [1]. Non-slip floor finish should be used throughout the drop-off area. Proper drainage should be maintained

(*Continued*)

(*Continued*)

Problem	Design	Construction	Maintenance
 Example of drop-off point for elderly and persons with disabilities	and BS 9266:2013). The buffer between the building entrance and car drop-off zone should have a minimum width of 2500 mm, and be provided with seating and guard rails. The drop-off points should be as close as possible to the main entrance, and should have a minimum length of 9000mm and a width of 3600mm, and be served by a kerb ramp (ISO/NP 21542).	Ground and floor surfaces along accessible routes and in accessible rooms and spaces — including floors, walks, ramps, stairs, and curb ramps — must be stable, firm, slip-resistant, and in compliance with the Code of Accessibility [11]. (See also BS 8300:2009+A1:2010 and BS 9266:2013).	to prevent ponding in drop-off areas [11]. Install signs to identify the drop-off zone, in order to prevent its misuse as a parking space [14]. (See also BS 8300:2009+A1:2010 and BS 9266:2013).
Insufficient outdoor Lighting  Example of a lighted outdoor arcade area  Example of an insufficiently lighted outdoor area	Adhere to the guidelines and design criteria for lighting of work places (outdoors) as per SS 531-2:2008(2014). (See also CIE S 015/E:2005). Provide a hierarchy of lighting effects that correspond to the different zones and uses of the outdoor area [16]. Ensure that lighting design will improve the energy and sustainability objectives of the building.	Building façade lighting fixtures must adhere to a total façade lighting power <5% of total interior lighting power (SS 530:2014). (See also ANSI/ASHRAE/IES Standard 90.1:2013). Ensure that display and ornamental lighting are separately controlled.	Comply with the minimum outdoor lighting requirements (SS 531-2:2008(2014) (See also CIE S 015/E:2005): • pedestrian walkways 5 lux • Slow moving traffic areas (<10km/h) 10 lux • Regular vehicular traffic (<40km/h) 20 lux • Pedestrian passages, loading/unloading points 30 lux • Reading labels 50 lux

(*Continued*)

(*Continued*)

Problem	Design	Construction	Maintenance
Slip and Fall  Example of a dry floor friction tester (FFT)  Example of a wet pendulum slip tester  Defective playground surface may cause tripping	Select playground surfacing material as per SS 495:2001, ASTM F 1292-04. Classify surface materials' slip resistance through Wet pendulum test, Dry floor friction test, Wet/ barefoot ramp test and Oil-wet ramp test. (See also BS EN 1177:2008). Conduct testing impact attenuation for a critical fall height for surfacing (SS 495:2001) (see also BS EN 1177:2008). Refer to AS/ NZS 4486.1: 1997 for guidelines for playgrounds' development, installation and maintenance.	The minimum slip resistance on surface systems (e.g. moulded tiles, mats, cast in-situ rubber surfaces) when tested for any direction, under either dry or wet conditions must not fall below 40 (SS 495:2001). (See also BS 7188:1998+A2:200). The use of Pendulum and Tortus methods for slip resistance tests are recommended (SS 485:2011). (See also AS 1683.15.1:2000; AS HB 197:1999 and AS/NZS 4663:2004). Refer to AS/NZS 4486.1:1997 for guidelines for playgrounds' development, installation and maintenance.	Minimum pendulum test classification of W for external walkways (SS 485:2011). (See also AS 1683.15.1:2000; AS HB 197:1999 and AS/ NZS 4663:2004). Keep footpath surfaces clear of any dirt, waste, or liquid spills (water) which might result in slip hazards [12]. Provide visual demarcation of wet floor areas and slip hazards [12] during cleaning and maintenance. Refer to AS/NZS 4486.1:1997 for guidelines for playgrounds' development, installation and maintenance.

5.2 Indoor Open Spaces

Problem	Design	Construction	Maintenance
Poorly-designed way-finding, directory and advertisement boards (issues with power supply and mounting details)  Directional signage finish reflects light and provides glare	Design work for wayfinding boards should begin with a comprehensive study of the environment, user groups and needs, and user-traffic flow (SS 599:2014). (See also ISO 28564-1:2010). Signage should be in a flexible modular system that is consist in design, materials, construction and finish.	Use sufficiently robust and durable materials with proper construction detailing to withstand the normal wear and tear of signage (SS 599:2014). (See also ISO 28564-1:2010). Refer to the minimum illuminance of all areas on signs — 200 lux or approximately 50 lux above the surrounding light level (whichever is	Provide adequate directional signage and provide for mounting, power and data points [1]. Allow efficient update of signage whenever information is subject to change [11]. Signage components should also be easy to clean, repair and update (SS 599:2014). (See also ISO 28564-1:2010).

(*Continued*)

(*Continued*)

Problem	Design	Construction	Maintenance
 Example of non-glare finish for accessibility signage  Example of fire exit sign located slightly above floor level to guide evacuation	Limit number of variations within the range of sign components in order to enjoy the benefits of economies of scale during production and maintenance [11]. Specify and ensure that signage face have glare-free finish. In Australia, letters and numbers on signs are required to be in Sans serif (14) or Helvetica medium fonts [15].	greater) for externally illuminated signage; 300 cd/m2 based on the white foreground of the backlit sign-face on internally illuminated (backlit) signs. For large-scale developments or where appropriate, signage strategies should preferably be tested in the form of mock-ups and be evaluated with a walkthrough as a user before final implementation. [11].	Formulate a comprehensive strategy for updating, maintenance, control and enforcement, future procurement, and records [11]. Allow a 200 lux level of illumination on signs when emergency lighting is used [15].
Poor ventilation  Example of a mini jet fan used to improve ventilation in an indoor carpark	Comply with the ventilation requirements of car park spaces (above ground/basement) using fan systems as per SS 553:2016. (See also ANSI/ASHRAE/IES Standard 90.1:2013). Use hot smoke test/CFD fire modelling to show effectiveness of jet fans system [17]. (See also AS 1668.2).	Jet fan system integrated with carpark fire safety system within the same level for smoke control. Minimum headroom for installation of jet fan system is 3m [17].	Compliance of air change rate with SS 553:2016. (See also ANSI/ ASHRAE/IES Standard 90.1:2013). Quarterly inspection for noise, accumulation of dust and controls function [18].
Insufficient/ compromised indoor Lighting  Example of an acceptably lighted hotel lift lobby	Factors affecting luminous environment (i.e. luminance distribution, illuminance, glare, directionality of light, colour aspect of the light and surfaces, flicker, daylight, maintenance) must be considered. Comply with the design valuesprovided in SS 531-1:2006(2013)	The lighting installation should meet the lighting requirements of a particular interior, task or activity without wasting energy. Recommended task lighting levels are provided as maintained illuminance (i.e. depends on maintenance characteristic of the lamp, the luminaire, the	The essential documents to be handed over to Maintenance are [19]: 1. As-built lighting layout 2. Lighting schedule 3. Lighting data sheets Lighting zoning in accordance to operational requirements. Use energy-efficient lighting

(*Continued*)

(*Continued*)

Problem	Design	Construction	Maintenance
 Example of an acceptably lighted common area  Example of an acceptably lighted basement carpark  Example of non-visible step demarcation light on escalator  Damaged exit sign; may interfere with egress during evacuation	Clause 5. (See also ISO 8995-1:2002). Ensure that the lighting in indoor spaces are efficient, comfortable, and safe for visual tasks in the work period; ensuring visual comfort, visual performance, and visual safety (SS 531-1: 2006(2013)). (See also ISO 8995-1:2002). For stairs, escalators, and travellators: maximum lighting power density is 6 W/m2 as per SS 530:2014. (See also ANSI/ASHRAE/IES Standard 90.1:2013). Ensure that common area receives >150% of designed illuminance on a typical rainy day. An automatic daylight control, either determined by a photo-sensor or schedule, is mandatory (SS 530:2014). (See also ANSI/ASHRAE/IES Standard 90.1:2013).	environment and maintenance programme). A sufficient level of lighting is required to maintain safety (ISO/TR 22411:2011). To reduce and eliminate stairway accidents, the switches that control the stair lights should be placed at a sufficient distance from the stairs — far enough to eliminate the risk of a person falling while reaching for the switch. Three-way switches should be installed at the top and bottom of the stairs [20]. Provide ceiling lights to orient people along walkways and use contrasting colour luminance at base boards, walls and doors to delineate access routes. especially for people with limited vision and people with autism or cognitive disabilities [15].	such as LED. All LED lightings must be visibly flicker-free when dimmed. Lux level and uniformity to comply with SS 531-1:2006(2013). (See also ISO 8995-1:2002). Where daylighting is available, lighting control strategies can be employed to reduce energy consumption while maintaining the desired illuminance level. Provide sensor to dim lights by 50% when no one is using the staircase (within compliance to local Fire Codes) (ANSI/ASHRAE/IES Standard 90.1-2016). Preparation of comprehensive maintenance schedule to include lamp replacement, luminaire cleaning intervals and cleaning method (SS 531-2:2008(2014)). (See also CIE S 015/E:2005).

(*Continued*)

(*Continued*)

Problem	Design	Construction	Maintenance
 Example of insufficient lighting at basement area	The lighting scheme should be designed with maintenance factor for selected lighting equipment, space environment and specified maintenance schedule, calculated — as defined in CIE 154:2003 (SS 531-2:2006 (2013).	Installation of lightings should be uniform; avoid extreme differences in the levels of brightness. Ensure that lights are acceptably bright but do not cause glare or cast shadows that would give rise to optical illusions [11].	Ensure that the illumination level for access routes is at 150 lux and that all ramps and staircases are well-illuminated [15].
Universal design of indoor open spaces  Ramp connecting to indoor area lacks grab bars/handrails to assist people using canes  Example of ramp with anti slip strips	Conform to the design guidelines for accessibility, horizontal/ vertical circulation, doors, staircases, lifts, clear headroom, accessible individual washrooms, seating spaces, drinking fountains, signage, etc. as per [11]. (See also BS 8300:2009+A1:2010 and BS 9266:2013). Key accessibility issues (including for management and maintenance of built environment) should be considered from early stages of planning (ISO/ NP 21542).	Comply with the construction requirements (e.g. plans, specifications, permits) for making buildings (residential/ shophouse/ office/shopping complex/hotel, etc.) accessible to persons with disabilities and families with young children [11]. (See also BS 8300:2009+A1:2010 and BS 9266:2013). Ensure that the grab bars and handrails installed will be able to resist a force of at least 1.3 kN; applied vertically or horizontally [11].	Implement proper housekeeping practices (e.g. 5S method: Sort, Set in order, Shine, Standardise, and Sustain) to prevent slips, trips, and falls; limit spills; ensure machine safety; prevent fires; and ensure that exits and access routes to fire equipment are clear [12]. Provide adequate janitor rooms with storage for equipment (such as trolleys, etc.). Provide power points for equipment such as vacuum cleaners.

5.3 Staircases

Problem	Design	Construction	Maintenance
General  Ponding at stairs during rainy days may cause users to slip and fall  Refuge area used for storage. Important note: refuge areas should not be used or occupied for purposes other than as a rescue point for persons with disabilities during evacuation [21]	Staircase design to comply with requirements (steps profile, detectable warning surfaces, stair handrails) set in the Code on accessibility [11]. (See also BS 8300:2009+A1:2010 and BS 9266:2013). A flight of stairs should have a minimum of 3 risers. As a safety precaution, flights containing only 1 or 2 steps should be avoided. After a maximum of 18 risers an intermediate landing should be provided. Floor landings must have a level platform of the same width as that of the stairs [11].	Risers should have a maximum dimension of 175mm and treads should have a minimum dimension of 275mm [13]. Risers and treads should have consistent dimensions. The minimum width of stairs should be 900mm and must be adjusted according to the expected flow of traffic [11]. Staircases of widths wider than 2300 mm should be separated by a handrail into segments between 1100mm and 1800mm [6]. Construction of safe holding areas or safe refuge/rescue assistance areas should comply with the fire safety and evacuation plan [15].	Building owner/operator must implement a regular maintenance programme for the staircase [22]. Periodic inspection as well as the rectification of any irregularities (e.g. accumulation of debris, damaged treads/risers/nosing, etc.), should be carried out by trained maintenance personnel. Ensure that escape staircases are adequately lit via an emergency power supply during emergencies [1]. Electric illumination sources must be kept in good operating condition [23].
Handrails  Staircase lacking handrails	Install handrails on both sides, between 800mm and 900mm high (from the pitch line vertically up to the top of the handrails), from the base of the stairs through the entire length. Extension beyond the top and bottom steps should not be less than 300mm in length [11]. (See also BS 8300:2009+A1:2010 and BS 9266:2013).	Handrail is most effective when height is approximately equal to the average height of the hip joint of users (ISO/TR 22411:2011). Additional handrails for children are recommended. Handrails for children should be provided at 600mm from pitch line [13].	Handrails and balustrades must be kept in good repair, firmly fixed, and structurally sound [23]. Ensure that the right grade of stainless steel is installed, to prevent corrosion.

(*Continued*)

(*Continued*)

Problem	Design	Construction	Maintenance
Slippery access routes  Resolved by installing anti-slip strips on the ramp (see picture)  Good practice: avoid incidents of users slipping/falling on wet routes by placing a detectable warning indicator on ramp when needed (e.g. after housekeeping works/rain)	Provide stair nosing based on minimum pendulum or ramp recommendations as per SS 485:2011, and install tactile strips at the start and end of every flight of stairs [1]. (See also AS 1683.15.1:2000; AS HB 197:1999 and AS/NZS 4663:2004). All steps should be fitted with non-slip nosing strips between 50mm and 65mm in width, with permanent contrasting colours. Nosing should have no abrupt undersides nor project more than 25mm over the back edge of the step [13]. (See also BS 8300:2009+A1:2010 and BS 9266:2013).	Ensure that nosing are securely fastened to the steps and are made of durable material (e.g. stainless steel, ceramic, cast-iron) that also suit the building's usage [1]. Provide visual contrast between landings and the top and bottom steps of a flight of stairs and ensure similar frictional characteristics between different materials used in steps and landings (ISO/NP 21542). Ensure that the constructed ramp is stable, firm and has slip resistant materials [15].	Install anti-slip strips along stairway steps to avoid slip-and-falls [11]. Dust, debris or spills on the floor can create a slipping hazard. Staircase surfaces must be cleaned regularly and kept dry and free of dirt and dust [22]. Install strips of a contrasting colour at the top and bottom of the ramp. Ensure that detectable warning indicators (e.g. signs, housekeeping equipment, etc.) are provided [15] especially during cleaning, maintenance and repair works.
Accessibility 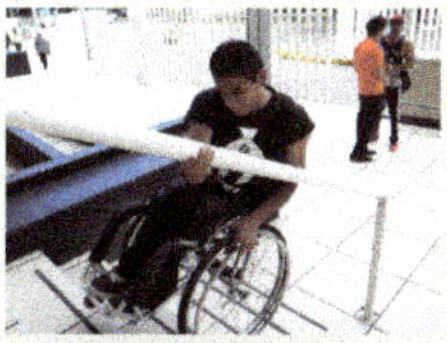 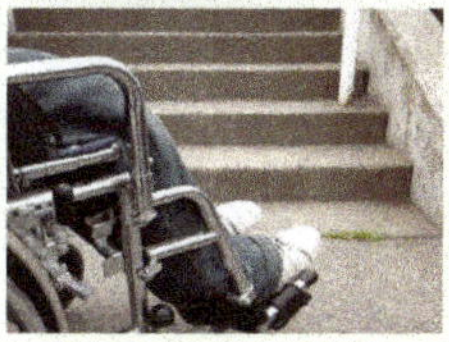 Lack of wheelchair access	Stairs and steps should be designed to accommodate the elderly and disabled people, and must comply with the provisions stipulated in the Code on accessibility in terms of considerations relevant to the design of stairs (ISO/TR 22411:2011). Provisions for staircases for ambulant disabled access include continuous handrails on both sides	The minimum clear headroom for all vertical circulation routes is 2000mm [6]. A detectable guardrail or other permanent barrier should be provided where headroom is less than 2000mm. Such elements should be at a maximum height of 580mm so that they can be detected by the visually impaired [11].	Building owner/operator must operate and maintain the required accessibility features in proper working condition for the use of any disabled personnel [13]. Alternative facilities or equipment need to be provided in case accessibility features are under repair or maintenance for a prolonged period of time [11].

(*Continued*)

(*Continued*)

Problem	Design	Construction	Maintenance
 Example of wheelchair lift	(extended >300mm beyond the top and bottom steps), tactile and Braille signs, uniform risers and no open risers [11]. (See also BS 8300:2009+A1:2010 and BS 9266:2013).	Design for wheelchair stair lifts should conform with provisions in the Code on accessibility [11]. (See also BS 8300:2009+A1:2010 and BS 9266:2013).	Maintenance of wheelchair lifts is important, to ensure the safety of passengers; thus only trained technicians should inspect, maintain and repair this equipment.

ACCESSIBILITY CONSIDERATIONS FOR ELEVATORS

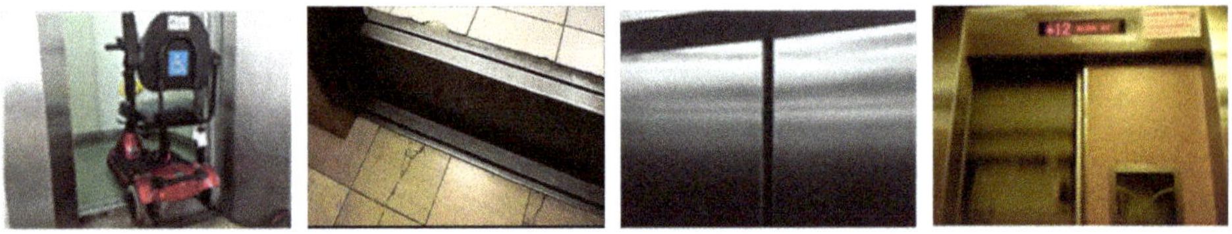

Examples of Elevator Defects (e.g. Elevator Car Landing Gap and Faulty Door Operation).

In designing and installing elevators, it is required that the internal car's dimensions allow moderately easy accessibility and manoeuvrability for persons in wheelchairs [24]. In circumstances where elevator size does not allow a person using a wheelchair to turn around, a mirror can be installed on the rear wall [15]. Designers are required to ensure that elevator controls are readily accessible to any user of diverse capabilities (e.g. 1200mm as the maximum height of the buttons; provision of Braille and tactile characters on the left of control panel button), as well as provide automatic verbal announcements that announce floor levels. Tactile and visual information be provided at a 45-degree angle from the wall so that standing people and people with visual impairments can read the characters. Install at least two sets of call buttons; one for people who are standing (higher set) and for people who are seated or of short stature (lower set). Provide a large illuminated floor indicator on a pad above the hall call buttons to indicate the elevator's location.

As shown in the above images, some elevator defects which cause harm and injury to passengers are due to faulty door operations and/or the elevator car landing gap. It is advisable that the following guidelines be considered during elevator repair and maintenance:

- Access to escalator or passenger conveyor shall be barred by suitable devices. Notices/signage such as "No access" or "No entry" must be provided (SS 626:2017).
- Adhere to the inspection criteria for safety of escalators as per JIS A 4302:2006. Upgrade the control system, braking, and motor types. Include a micro-processor controller that will electronically monitor and control motor rotation to ensure that the elevator car accurately stops at floor level [25]. (See also ASME A18.1-1999, ASME A18.1 (2005) and ASME A17.1-2000).
- Review levelling of car to ensure the value is acceptable by standards to avoid risk of passengers tripping and falling (e.g. wheelchair users) [26]. Always ensure that the Permit to operate (PTO) is displayed in the lift(s) [27].

(*Continued*)

(*Continued*)

- Inspect elevator door and guides, shoes, and tracks. They should not show any permanent deformation and elastic deformation greater than 15mm (when force >300 N is applied to area of 5cm^2 at the centre of the door panels at a right angle) (SS 550:2009). (See also ASME A18.1-1999, ASME A18.1-2003 and ASME A17.1-2000).
- Review the service call frequency for issues pertaining to the door, since increased service calls signify the need to upgrade/replace door operators [27].

Additional design, construction and maintenance guidelines pertaining to the mechanical and electrical systems for elevators are presented in Chapter 6.

5.4 Landscaping

Problem	Design	Construction	Maintenance
Green Roof/Green Wall issues  Dying plants on a green roof  Biological growth on a green wall  Root penetration through the wall structure	The structural integrity of a green wall system should be certified by professional structural engineer. Safety and maintenance concerns should be thought of and addressed early, during the design stage [28]. Frequency of maintenance and cost issues should be considered when selecting the plants and materials [29]. Considerations for a safe environment rely mostly on matching the proper plant to the proper place [16]. New gutters for green roofs should be provided with stainless steel leaf covers, and should be located at or less than 15mm above ground level (SS 525:2006) (See also BS EN 1253-2:2015).	The constructed roof garden should not allow plant roots to damage the building structure (e.g. roof waterproofing membrane) [30]. Safe work schedules to be followed with proper equipment and site supervision. Especially to prevent falls from heights and falling objects [29]. Inspect plants carefully upon their arrival at the site for damage to leaves/stems and broken branches during transport and handling. The landscape contractor should refer to the plant list for seasonal requirements related to the time of planting, and to contract specifications for additional requirements [16].	Prevent obstruction of structural elements to ensure easy periodic maintenance. Conduct periodic inspection of supporting structures [31]. Hardy plants that require less water and sunlight (especially for those installed in indoor environments) and routine maintenance practices are required in roof garden maintenance [32]. Monitor conditions on a regular basis to head off problems such as insect attacks, irrigation, disease or soil problems. Refer to the ANSI standards for pruning (ANSI A300:2017); and ANSI safety requirements for tree care operations (ANSI Z133:2012) [16].

(*Continued*)

(Continued)

Problem	Design	Construction	Maintenance
 Water seepage on column  Vegetation planted at areas that are difficult to access for maintenance  Access to vertical greenery is hampered by the height of the walkway ledge, where maintenance crew will need to climb onto it.	Adhere to proper specifications in waterproofing (SS 374:1994) to prevent water seepage on columns between two vertical greenery systems. (See also BS EN 13416:2001 and BS EN 12039:2016). Inaccessible or difficult to access green roof/green wall installations create issues in maintenance planning and execution. It has implications for labour costs such as increased man hours and the resultant higher cost. They also pose safety risks triggered by the lack of access and safety provisions [33].	Use of sharp and extremely angular grained aggregates, which can affect the stability of material during installation, is discouraged. Installation of staking and guying can take reference from the methods and standards specified in the latest ANSI A300 Part 3 and should be read in conjunction with the rest of the ANSI A300 when carrying out such installations (ANSI A300:2017). Maintain construction quality control during the installation of green roof and green wall component (e.g. fixtures and fittings and vegetation plantings).	Rainwater pipe works, gutters and gratings on green roofs should be inspected and thoroughly cleaned annually (SS 525:2006). The drip irrigation piping should preferably have provisions to counter plant-room penetration as plant roots may clog up drip holes [33]. Conduct annual or semi-annual inspections to control or prevent bird interaction with vertical greenery system (VGS), for plant survival must be taken into account. To minimise evaporation losses, water the plants between 4–7am or 6–9pm [34].
Chokage of discharge  Water stagnation due to chokage (at outdoor landscape)	New gutters should be provided with stainless steel leaf covers, and should be located at or less than 15mm above ground level (SS 525:2006). (See also BS EN 1253-2:2015). Locate drains at the edge of landscaped areas [16].	Provide eave gutters to maintain fall of 1 in 150 to prevent water ponding and debris build-up. Roof covering should not extend into the gutter, so as to accommodate easy cleaning and maintenance (SS 525:2006). (See also BS EN 1253-2:2015).	Inspect and thoroughly clean rainwater pipe works, gutters and gratings annually (SS 525:2006). (See also BS 7370-5:1998). Monitor conditions on a regular basis to head off problems such as insect attacks, disease, or soil problems.

(Continued)

(*Continued*)

Problem	Design	Construction	Maintenance
 Water stagnation due to chokage (at roof garden)	Drain inlets should be accessible, durable and appropriate for the climate zone [16]. Adhere to the landscape's specifications for proper drainage layout and system as per BS 4428:1989.	Maintain construction quality control during the installation of green roof/ green wall components, waterproofing, pipes and fittings, and the planting of vegetation.	Long term care requires periodic cleaning and sealing of joints and re-setting of grates and pavement finishes or turf grades due to settlement or silting [16].

5.5 Swimming Pools

For additional guidelines concerning ancillary services/systems (e.g. heating, plumbing, drainage, ventilation and lighting systems) necessary for the operation and maintenance of swimming pools, refer to Chapter 6: Mechanical, Electrical and Plumbing Systems.

Problem	Design	Construction	Maintenance
Slip and Fall  Warped wooden flooring. A common cause of trip-and-fall incidents	Design considerations include; amount and type of expected traffic; wear resistance and cleanability of material; exposure to anticipated contaminants; and environmental factors (e.g. visibility issues and contamination minimisation)	Conduct tests for slipperiness using wet barefoot ramp test method and wet pendulum test method as per SS 485:2011 (See also AS HB 197:1999 and AS/NZS 4663:2004). Conform to the recommended	Conduct regular high jetting of perimeter to prevent moss growth due to dampness. Ensure that inspections and tests are carried out at the specified intervals as a preventative measure, and that any remedial action required is promptly dealt with.

(*Continued*)

(Continued)

Problem	Design	Construction	Maintenance
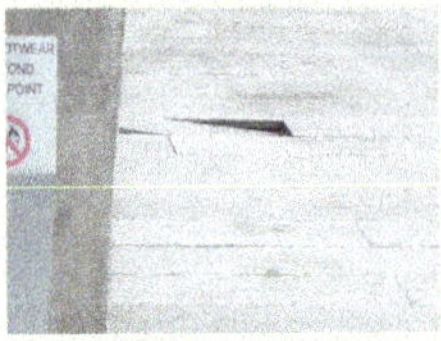 Broken wooden tiles. Another slip-and-fall hazard to pool users  Good practice: Use non slippery pool tiles (example in picture)	(SS 556:2010). (See also AS 2610.1-2007 and AS 2610.2-2007). Use non-slippery and smooth grip tiles along the edges of the pools [9, 10]. Abrupt changes in floor level, including steps, should be avoided in 'wet' areas wherever possible. The slip resistance of any given surface will diminish if the gradient becomes steeper than 1 in 30 or is less than 1 in 60 [23]	minimum Pendulum test values for pedestrian flooring in the pool's surroundings, communal shower rooms (W), pool ramps, and stairs leading into water (V) (SS 485:2011). (See also AS HB 197:1999 and AS/NZS 4663:2004). During construction, ensure that there is a wide band of tile of a contrasting colour near the edge of the pool, which slightly slopes upwards to indicate to people (especially people with limited vision) that they are nearing the edge of the pool [15].	Ensure that the floors and stairs are kept clean, are drained where necessary, and are not slippery [10]. Comply with the maintenance policy and practices in compliance with occupational, health and safety requirements; special provisions for slip hazards (guards and handrails); and alternative information sources (use of contrasting colours and warning signs) (SS 485:2011). (See also AS HB 197:1999 and AS/NZS 4663:2004).
Compromised/poor access and egress  Crack on fibreglass pool stairs [35]  Good practice: provide ramp for people with disabilities (example in picture)	Provide access and egress from pool using one or a combination of stairs/steps, ladders, swim-outs, pool-seats, landings, ramps or beach entries (SS 556:2010). (See also BS EN 16582-1:2015 and BS EN 16582-2:2015). Ramps may be provided for people with disabilities, for ease of access to the pool. It should have a gradient that does not exceed 1 in 15; have a clear 1m width; have a slip-resistant surface and handrails on both sides of the ramp [23].	The construction should bear full and empty conditions of the pool. Pool features should be constructed with inert, stable, non-toxic materials that are durable and watertight [23]. Ensure that during and after construction, the facility is free from any obstructions which might cause entrapment or injury [9]. Ensure that floor and stair surfaces are durable and slip-resistant [9, 10].	Specify and ensure that the right grade of stainless steel is installed, to prevent corrosion. Ensure that inspections are carried out at the specified intervals as a preventative measure, and any remedial action that is required is promptly dealt with. Ensure that the floors and stairs are kept clean, are drained where necessary, and are not slippery [10].

(Continued)

(*Continued*)

Problem	Design	Construction	Maintenance
Building materials  Corrosion of column and tiles along pool stairs  Defective and warped wooden flooring by the pool causes water ponding and slipping  Safety signage and behaviour rules are located away from the entrance, and cannot be directly seen by all users  Stains on pool tiles	Ladders must be made of corrosion-resistant materials and have slip-resistant tread surfaces with maximum distance (300mm) between ladder rungs. Provide two handholds or handrails to all rungs (max. dia. of handrail 50mm and min. dia. of handrail 30mm) (SS 556:2010). (See also AS 2610.1-2007 and AS 2610.2-2007). Consider water presence, pool, and cleaning chemicals when specifying building materials (e.g. concrete, tiles, adhesives, grouts) (BS EN 1992-3:2006). (See also AS 2610.1-2007 and AS 2610.2-2007). Floor gullies, gutters and valleys should not constitute a tripping hazard; refer to the detailed design issues related to these features as per Time Saver Standards [23]	The construction should bear full and empty conditions of the pool. The pool/pool features should be constructed with inert, stable, non-toxic materials which are durable and watertight. Working stresses are to be based on predetermined ultimate strengths of materials used, with a factor of safety of not less than 2t/z [23]. (See also BS EN 1992-1:2006, BS EN 1992-3:2006, BS EN 15288-1:2008, AS 3600-2009 and AS 3735-2001). Appropriate surface finishes should be used. They are to enhance the safety and hygiene of the premises, and to assist in effective maintenance by enabling dirt and visible contaminants to be easily detected [10]. Walls and floors should be smooth, impervious, durable, easily cleaned and continuous, with no cracks, joints or protrusions other than structural joints (SS 556:2010). (See also AS 2610.1-2007 and AS 2610.2-2007).	Materials used for landscaping of the pool edge should have a smooth surface, to facilitate easy cleaning (SS 556:2010). (See also AS 2610.1-2007 and AS 2610.2-2007). Ensure the clear display of safety signage and behaviour rules [10]. (See also ISO 20712-1:2008 and AS/NZS 2416.3:2010). Ensure that the concourse directly surrounding aquatic facilities does not accommodate considerable amounts of water (to avoid pooling and microbial growth), nor have any irregularities (to avoid slip-and-trip hazards) (SS 556:2010). (See also AS 2610.1-2007, AS 2610.2-2007 and BS EN 15288-2:2008). Where divers are used for installing, maintaining, repairing or cleaning of swimming pools, the requirements of the Diving at Work Regulations should be followed [10].

(*Continued*)

(*Continued*)

Problem	Design	Construction	Maintenance
Water quality — pH Value  Traces of dirt and algae growth at the tile rim of planter due to close proximity of plant landscape to the pool  Algae growth turns the pool green and swampy [36] 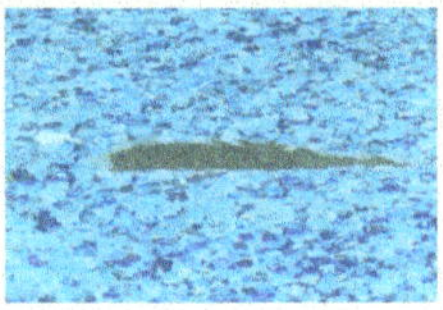 Soil and green patches from the damaged pool flooring	Provision of accurate and reliable testing kit(s) (to measure the pool water's pH level, as well as to check for free residual chlorine, total available chlorine, bromine or other chemical disinfectant residuals, cyanuric acid (if used), total alkalinity, calcium hardness, and copper and silver. Daily records must be maintained (ISO 17381:2003). The pool must be so designed that the water quality will always remain safe for the public during its operation [37]. Refer also to BS EN 16713-1: 2016; BS EN 16713-2: 2016; BS EN 16713-3: 2016 and AS/NZ 1926.3-2010. Ensure that pool design specifications require the maintenance of the pool water at a pH value between 7.2 and 7.8 [37].	Do not plant landscape and trees too close to pool, to prevent overflow of water from the planting strip(s) that may contaminate the water in the pool and also create maintenance problems [37]. The choice of material for pool furniture should withstand dampness and exposures to UV light. Ensure that during and after construction, there are no depressions in the concourse which will result in water pooling and/or lead to microorganism growth [9, 10]. Installation of the water treatment system should conform as per SS 556:2010. Refer also to BS EN 16713-1:2016; BS EN 16713-2:2016; BS EN 16713-3:2016 and AS/NZ 1926.3-2010.	Use adequate level of a chemical in water to destroy micro-organisms (e.g. chlorine). Reduce chloramines in water using ozone, UV light irradiation, or addition of non-chlorine oxidising chemicals. (SS 556:2010). (See also BS EN 16713-3:2016 and AS/NZ 1926.3-2010). Conduct monthly bacteriological sampling and constantly check that the disinfectant level and pH value are correct, to ensure the bacteriological quality of a well-run pool [10]. Water quality testing should be conducted by an accredited laboratory at least monthly for physico-chemical parameters (SS 556:2010, BS EN 16713-3:2016). Tests for microbiological analysis should also be carried out (ISO 19458:2006).

(*Continued*)

(*Continued*)

Problem	Design	Construction	Maintenance
Sharp edges due to breakage of tiles  Tile breakage at pool sides  Chipping of tiles pool's sides	Aquatic facility water bodies should not be designed or constructed with obstructions that can cause users to become trapped or injured (SS 556:2010). (See also BS EN 16582-1:2015 and BS EN 16582-2: 2015). Fixtures and fittings in the walls and floors of the water body shall be fitted flush and have no sharp and protruding edges. Specify that all the edges and corners of the facility shall be rounded [9, 10].	Tile selection during construction should be based on tile properties (safety, slip-resistance, weight, chemical and abrasion resistance, water absorptivity) and traffic considerations [9]. Ensure that all the edges and corners of the facility are of round finish [9, 10]. Ensure that the expansion joints created between the coping and the pool deck are caulked.	Conduct routine safety audits on existing pools to improve facility safety; e.g. by replacing/overlaying finishes with anti-slip finishes and providing additional safety features to address protrusions in the pool area (SS 556:2010). (See also BS EN 16582-1: 2015 and BS EN 16582-2:2015). Ensure that inspections are carried out at the specified intervals as a preventative measure, and that any remedial action required is promptly dealt with [10].

DESIGN CONSIDERATIONS FOR POOL ANCILLARY SERVICES

The swimming pool defects shown above could be caused by the design failure of the swimming pool's ancillary services. Photo 1 shows a defective sand filtration pump where sand accumulates on the pool's surface, while Photo 2 shows a water recreational area at the roof deck, which has high maintenance and operation costs. Photo 3 shows a condominium's infinity pool's shattered glass panels and the water falling from the fifth floor in consequence — the incident was caused by a blocked drain [38]. All these defects, which are considered environmental and safety hazards, can be prevented — if properly designed, and the requirements set by the mentioned list of standards for the construction and maintenance of swimming pools are adhered to.

Based on the Code of Practice on Environmental Health [37], additional design criteria are specified to address environmental health and safety concerns in the design of swimming pools such as (a) provision of a water-

(*Continued*)

(*Continued*)

circulation system to facilitate complete circulation of the water through all parts of the pool; (b) provision of overflow weirs for at least 50% of pool's perimeter (a deck level channel design can be adopted for the overflow weir drainage system for ease of maintenance); (c) provision of at least one standby pump unit and motor to supplement the duty pump provided in the filtration system; (d) installation of flow meters capable of measuring water flows of 1.5 times the designed flow rate on all re-circulation systems; (e) ensuring that the filtration plant adopts rapid sand, diatomaceous earth or any other acceptable filtration system; (f) provision of sampling taps at the inlet and outlet pipes of the filter; (g) ensuring that the filter backwash water is discharged into the sewer via a backwash water holding tank; (h) supplying the pool with automatic disinfectant and chemical feeders to maintain the bacteriological and chemical characteristics of the water within the water quality limits; (i) ensuring that landscaping materials used have smooth surfaces to facilitate easy cleaning; (j) If a submerged bar is provided in the pool, provision of a sink connected to a sewer must be provided for the installed submerged pool bar; (k) provision of at least two showers around the swimming pool; and (l) provision of adequate ventilation in the case of indoor pools.

Additional guidelines for the design and installation of pool ancillary services and features are highlighted as follows:

- Installation of mechanical ventilation systems for indoor aquatic facilities should be adequate as well as adhere to the provisions of AS 1668.2. (See also SS 553:2016 and BS EN 1886:2007).
- Sufficient lighting should be provided to swimming pools while in use. Ensure that there is no direct interference or reflected glare from lighting sources. Appropriate lighting should also be provided for indoor swimming pools.
- Installation of shade structures for all outdoor swimming facilities is recommended, provided that their location does not obstruct overhead lighting [10].
- All electrical installations should conform to AS 3000 "Wiring Rules". (Refer also to AS/NZS 3000:2007, AS/NZS 60598.2.18:1998, AS/NZS 3136:2001, BS 7671:2008+A3:2015 and IEC 60335-2-60 Ed 3.1).
- Lightning protection systems should be installed in Group 1 (public access with limited restrictions, such as age without an accompanied adult) and Group 2 (restricted to discrete users and user groups) swimming facilities. They should adhere to the provisions of AS 1768. (See also BS EN 62305-1:2011).
- When installing solar water heating systems, ensure that the construction materials used do not contaminate water, and are not susceptible to corrosion under normal service conditions. The installation of a temperature control system will ensure pool users will not be exposed to water temperatures exceeding 38°C. Solar pool heating systems must be installed on a plumbing circuit that is separate and independent from the filtration system [10].
- An automatic or manual drainage system must be installed to enable the emptying of all the water from system when not in use. The drainage system must incorporate a back-flow prevention valve to prevent water from flowing back into the pool through the filter [10].
- Filtration equipment, outlet devices and skimmer boxes used for pools and spas should be designed and installed by adhering to the provisions in AS 2610.2-2007. (See also BS EN 16713-1:2016 and BS EN 16713-3:2016).
- Plumbing products and fittings for swimming pools and spas should be installed in accordance to the specifications included in AS/NZS 3500.1:2003 and AS/NZS 3500.4:2003. (Refer also to AS/NZS 3500 (Set):2003 and BS EN 16713-2:2016).
- Playground equipment (including water slides) used near or with pools and spas should be designed by complying with the requirements specified in AS 4685. Detailed information on the manufacturing and design for all types of slides is included in AS 4685.3-2004. See also AS/NZS 4422:1996.
- All areas of waterslide facilities that are available to the public must also comply with the requirements included in AS 4685.3-2004, AS 2560.2.5-2007 and AS/NZS 60598.2.18:1998. (See also BS EN 13451-5:2014, BS EN 1069-1 and BS EN 1069-2).

References

[1] Building and Construction Authority (2007). *Universal Design Guidelines*. Singapore: BCA.

[2] National Disability Authority (2017). Building for Everyone: A Universal Design Approach. Retrieved on March 22 from http://universaldesign.ie/ Built-Environment/ Building-for-Everyone/Entire-Series-Books-1_10.pdf.

[3] Kellet, J. and Rofe, M. (2009). Creating Active Communities: How Can Open and Public Spaces in Urban and Suburban Environments Support Active Living?: A Literature Review. Australia, Institute for Sustainable Systems and Technologies, University of South Australia to SA Active Living Coalition.

[4] Campbell, J. W. P. and Tutton, M. eds. (2014). *Staircases: History, Repair and Conservation. Campbell*, New York: Routledge.

[5] Templer, J. A. (1994). *The staircase: studies of hazards, falls and safer design*. USA: Massachusetts Institute of Technology.

[6] Singapore Civil Defence Force (2013). *Code of Practice for Fire Precautions in Buildings*. Singapore: SCDF.

[7] Landscape Network (2017). What is landscaping? Retrieved on March 22 from https:// www.landscapingnetwork.com/landscape-design/what-is.html.

[8] Baker, M. (2005). *The swimming pool: stylish and inspirational ideas for building and decorating your pool*. New York: Clarkson Potter.

[9] Alliance for Water Efficiency (2017). Swimming Pool and Spa Introduction. Retrieved on March 22 from http://www.allianceforwaterefficiency.org/Swimming_Pool_and_ Spa_ Introduction.aspx.

[10] Health and Safety Executive (2003). *HSG179: Managing health and safety in swimming pools* (3rd ed.). England: HSE Books and Sport England Publications.

[11] Building and Construction Authority (2013). *Code on accessibility in the built environment*. Singapore: BCA.

[12] Workplace Safety and Health Council (2016). *Workplace Safety and Health Guidelines on Workplace Housekeeping*. Singapore: WSHC.

[13] Building and Construction Authority (2002). *Code on Barrier-Free Accessibility in Buildings (BFA)*. Singapore: BCA.

[14] United Nations (2017). Accessibility for the Disabled — A Design Manual for a Barrier Free Environment: Urban Design Considerations. Retrieved on March 22 from http://www.un.org/esa/socdev/enable/designm/AD1-07.htm.

[15] Canadian Human Rights Commission (2007). *International Best Practices in Universal Design: A Global Review* (Rev. ed.). Canada: Canadian Human Rights Commission.

[16] Harris, C.W. and Dines, N.T. (1998). *Time Saver Standards for Landscape Architecture: Design and Construction Data* (2nd ed.). USA: McGraw-Hill.

[17] Singapore Civil Defence Force (2008). *Fire safety requirements for ductless jet fans system in car parks*. Singapore: SCDF.

[18] Chew, M. Y. L. (2016). *Maintainability of Facilities: Green FM for Building Professionals* (2nd ed.). Singapore: World Scientific.

[19] Building and Construction Authority (2016). *Green Mark for Non Residential Buildings NRB: 2015*. Singapore: BCA.

[20] Watson, E. and Crosbie, M. J. (2004). *Time Saver Standards for Architectural Design: Technical Data for Professional Practice* (8th ed.). USA: McGraw-Hill Professional.

[21] Photo by SOJIBSAMS (2016). Innovation Technology on Fire Safety. Retrieved on May 22 from https://www.slideshare.net/SOJIBSAMS/innovation-technology-on-fire-safety

[22] Singapore Civil Defence Force (2008). *Operation & maintenance manual for staircase storey shelters*. Singapore: SCDF.

[23] De Chiara, J. and Crosbie, M.J. (2001). *Time Saver Standards for Building Types* (4th ed.). USA: McGraw-Hill.

[24] Office of Housing and Construction Standards (2014). *The Building Access Handbook Building Requirements for Persons with Disabilities from British Columbia Building Code 2012 including Illustrations and Commentary*. British Columbia: Office of Housing and Construction Standards.

[25] Balmer, D. (2015). Single Speed Elevators: Time to Retire. Retrieved on March 22, 2017 from http://www.skyline-elevator.com/wp-content/uploads/2010/09/SingleSpeedElevators.pdf.

[26] Building and Construction Authority (2016). Building Maintenance and Strata Management Act. (Lift, Escalator and Building Maintenance) Regulations 2016. Singapore: BCA.

[27] Piper, J. (2006). Avoiding Elevator Breakdowns. Retrieved on March 22, 2017 from http://www.facilitiesnet.com/elevators/article/Keeping-Up-To-Avoid-Going-Down-Facilities-Management-Elevators-Feature — 4810.

[28] Centre for Urban Greenery and Ecology (2014). *Guidelines on Design for Safety of Skyrise Greenery*. Singapore: National Parks Board.

[29] Centre for Urban Greenery and Ecology (2013). *A Concise Guide to Safe Practices for Rooftop Greenery*. Singapore: National Parks Board.

[30] Centre for Urban Greenery and Ecology (2012). *Guidelines on Waterproofing for Rooftop Greenery*. Singapore: National Parks Board.

[31] Centre for Urban Greenery and Ecology (2013). *A Concise Guide to Safe Practices for Vertical Greenery*. Singapore: National Parks Board.

[32] Centre for Urban Greenery and Ecology (2012). *Guidelines on General Maintenance for Rooftop Greenery*. Singapore: National Parks Board.

[33] NUS Maintainability of Buildings (2017). Roof/Sky Garden. Retrieved on May 9 from http://www.hpbc.bdg.nus.edu.sg/?page_id=963.

[34] NUS Maintainability of Buildings (2017). Vertical Greenery. Retrieved on May 9 from http://www. hpbc.bdg.nus.edu.sg/?page_id=962.

[35] Photo by UV Pools (2017). Fiberglass Pool Staircase Repair. Retrieved on March 22 from https://uvpools.com/fiberglass-pool-staircase-repair/.

[36] Photo by Tom Daley taken from Menezes, J. D. (2016). Rio 2016: Green Olympic pool drained as organisers reveal hydrogen peroxide caused colour change. Retrieved on June 1, 2017 from http://www.independent.co.uk/sport/olympics/rio-2016-green-

pool-drained-as-olympic-organisers-reveal-hydrogen-peroxide-is-cause-of-colour-change-a7189616.html.

[37] National Environment Agency (2005). *Code of Practice on Environmental Health*. Singapore: National Environment Agency (NEA).

[38] The Strait Times (2016). Blocked drain caused condo glass panels to shatter. Retrieved on June 1, 2017 from http://www.straitstimes.com/singapore/housing/blocked-drain-caused-condo-glass-panels-to-shatter.

Normative References/Standards Referred to for Common Area

- ANSI A300:2017 — Tree, Shrub, and Other Woody Plant Management — Standard Practices (Pruning)
- ANSI Z133-2012 — Safety Requirements for Arboricultural Operations
- ANSI/ASHRAE/IES Standard 90.1:2013 — Energy standard for buildings except low-rise residential buildings
- AS 1668.2-2012 — The use of ventilation and airconditioning in buildings — Mechanical ventilation in buildings
- AS 1683.15.1:2000 — Methods of test for elastomers — International rubber hardness
- AS 1768 — Lightning Protection
- AS 1926.3-2010 — Swimming pool safety — Water recirculation systems
- AS 2560.2.5-2007 — Sports lighting — Specific applications: Swimming pools
- AS 2610.1-2007 — Spa pools — Public spas
- AS 2610.2-2007 — Spa pools — Private spas
- AS 3600-2009 — Concrete structures.
- AS 3735-2001 — Concrete structures for retaining liquids
- AS 4685.1-2004 — Playground equipment — General safety requirements and test methods
- AS 4685.3-2004 — Playground equipment — Particular safety requirements and test methods for slides.
- AS HB 197:1999 — An introductory guide to the slip resistance of pedestrian surface
- AS/NZS 4486.1:1997 — Playgrounds and playground equipment — Development, installation, inspection, maintenance and operation
- AS/NZ 1926.3-2010 — Swimming pool safety — Part 3: Water recirculation systems
- AS/NZS 2416.3:2010 — Water safety signs and beach safety flags — Guidance for use
- AS/NZS 3000:2007 — Electrical installations (known as the Australian/New Zealand Wiring Rules).
- AS/NZS 3136:2001 — Approval and test specification — Electrical equipment for spa and swimming pools
- AS/NZS 3500 (Set):2003 — Plumbing and drainage set
- AS/NZS 3500.1:2003 — Plumbing and drainage — Water services
- AS/NZS 3500.4:2003 — Plumbing and drainage — Heated water services

- AS/NZS 4422:1996 — Playground surfacing — Specifications, requirements and test method
- AS/NZS 4663: 2004 — Slip resistance measurement of existing pedestrian surfaces
- AS/NZS 60335.2.60:2006 — Household and similar electrical appliances — Safety — Particular requirements for whirlpool baths and whirlpool spas
- AS/NZS 60598.2.18:1998 — Luminaires Particular requirements — Luminaires for swimming pools and similar applications
- ASME A17.1-2000 — Safety Code for Elevators and Escalators
- ASME A18.1-1999 and ASME A18.1-1999 — Safety Standard for Platform Lifts and Stairway Chairlifts
- ASME A18.1 (2005) — Safety Standard for Platform Lifts and Stairway Chairlifts
- ASTM F 1292-04 — Standard Specification for Impact Attenuation of Surfacing Materials Within the Use Zone of Playground Equipment
- BS 4428:1989 — Code of practice for general landscape operations (excluding hard surfaces)
- BS 7188:1998+A2:200 — Impact absorbing playground surfacing. Performance requirements and test methods
- BS 7370-5:1998 — Grounds maintenance. Recommendations for the maintenance of water areas
- BS 7671:2008+A3:2015 — Requirements for Electrical Installations. IET Wiring Regulations.
- BS 8300:2009+A1:2010 — Design of buildings and their approaches to meet the needs of disabled people. Code of practice.
- BS 9266:2013 — Design of accessible and adaptable general needs housing. Code of practice
- BS EN 1069-1 — Water slides. Part 1. Safety requirements and test methods.
- BS EN 1069-2 — Water slides. Part 2. Instructions.
- BS EN 1177:2008 — Impact attenuating playground surfacing. Determination of critical fall height
- BS EN 12039:2016 — Flexible sheets for waterproofing. Bitumen sheets for roof waterproofing. Determination of adhesion of granules
- BS EN 1253-2:2015 — Gullies for buildings. Roof drains and floor gullies without trap
- BS EN 13416:2001 — Flexible sheets for waterproofing. Bitumen, plastic and rubber sheets for roof waterproofing. Rules for sampling
- BS EN 13451-5:2014 — Swimming pool equipment. Additional specific safety requirements and test methods for lane lines and dividing line
- BS EN 15288-2:2008 — Swimming pools. Safety requirements for operation
- BS EN 16582-1:2015 — Domestic swimming pools. General requirements including safety and test methods
- BS EN 16582-2:2015 — Domestic swimming pools. Specific requirements including safety and test methods for inground pools
- BS EN 16713-1:2016 — Domestic swimming pools. Water systems. Filtration systems. Requirements and test methods
- BS EN 16713-2:2016 — Domestic swimming pools. Water systems. Circulation systems. Requirements and test methods

- BS EN 16713-3:2016 — Domestic swimming pools. Water systems. Water treatment. Requirements
- BS EN 1886:2007 — Ventilation for buildings. Air handling units. Mechanical performance
- BS EN 1992-1:2006 — Design of concrete structures. General rules and rules for buildings
- BS EN 1992-3:2006 — Design of concrete structures — Part 3: Liquid retaining and containment structures
- BS EN 62305-1:2011 — Protection against lightning. General principles
- CIE S 015/E:2005 — Lighting of Outdoor Workplaces
- CIE 154:2003 — The Maintenance of Outdoor Lighting Systems
- IEC 60335-2-60 Ed 3.1 — Household and similar electrical appliances — Safety — Part 2-60: Particular requirements for whirlpool baths and whirlpool spas
- ISO 17381:2003 — Water quality — Selection and application of ready-to-use test kit methods in water analysis
- ISO 19458:2006 — Water quality — Sampling for microbiological analysis
- ISO 20712-1:2008 — Water safety signs and beach safety flags — Part 1: Specifications for water safety signs used in workplaces and public areas
- ISO 28564-1:2010 — Public information guidance systems — Part 1: Design principles and elements requirements for locations plans, maps and diagrams
- ISO 8995-1:2002 — Lighting of work places — Part 1: Indoor
- ISO/IEC Guide 71:2014 — Guide for addressing accessibility in standards
- ISO/NP 21542 — Building construction — Accessibility and usability of the built environment
- ISO/TR 22411:2011 — Ergonomics data and guidelines for the application of ISO/IEC Guide 71 to products and services to address the needs of older persons and persons with disabilities
- JIS A4302:2006 — Inspection Standard of Elevator, Escalator and Dumbwaiter
- SS 485:2011 — Specification for slip resistance classification of pedestrian surface materials
- SS 495:2001 — Impact attenuation of surface systems under and around playground equipment
- SS 525:2006 — Code of practice for drainage of roofs
- SS 530:2014 — Code of practice for energy efficiency standard for building services and equipment
- SS 531-1:2006(2013) — Code of practice for lighting of work places — Indoor
- SS 531-2:2008(2014) — Code of practice for lighting of work places — Outdoor
- SS 553:2016 — Code of practice for air-conditioning and mechanical ventilation in buildings
- SS 556:2010 — Code of practice for the design and management of aquatic facilities
- SS 599:2014 — Guide for wayfinding signage in public areas
- SS 626:2017 — Code of practice for design, installation and maintenance of escalators and moving walks
- SS 550:2009 — Code of practice for installation, operation and maintenance of electric passenger and goods lifts

Chapter

Mechanical and Electrical Systems

Introduction

Mechanical, electrical and plumbing systems (MEP systems) are any building services that use machines to ensure that buildings operate well, and that promote occupant health, comfort and productivity [1].

MEP systems contribute to the building's sustainability and energy demand, which significantly affect the building's cost [2]. The MEP systems in this chapter refers to the HVAC (heating, ventilation and air conditioning) systems, plumbing-sanitary systems, electrical systems, fire protection systems and mechanical transportation systems (e.g. elevators, escalators and moving walkways). How well the building is designed with maintenance considerations will significantly impact the building's performance. Faulty design of MEP services can be accounted to the lack of consideration for the design for access, equipment availability and maintenance requirements. Incorporating specific provisions for undertaking inspections and maintenance tasks in the design and planning stage will provide ease in performing inspections on and maintenance of MEP systems. To reduce the repair time of the system during operation stage, the recommendation and implementation of standardised parts and assemblies can be done during the design stage.

Designing MEP systems entails understanding the surrounding environment and probing for a simple and effective approach. where the resulting system is the best fit for the building, and the causes of failure can be easily detected and quickly repaired.

View of an AHU room.

Mechanical and electrical installation (air handling unit) at a commercial building.

This design approach will aid in minimising redundancy and failure rates. Considerations for the maintainability of the MEP systems during the design stage is vital for achieving: (a) extended systems and components' lifespan, (b) reduced operations cost by functioning at high productivity, (c) reduced unplanned system interruptions by avoiding failures, and (d) lesser unplanned system stoppages by reducing the expected time to perform a specific maintenance job [3]. This chapter highlights design, construction and maintenance guidelines for MEP systems that provide relevant specifications that address each prevalent defect of MEP systems.

Modern HVAC systems are designed not only to meet cooling and relative humidity requirements, but to also provide equipment efficiency, envelope construction methods, and optimise system demands, in order to achieve high energy efficiency. Being a complex system, HVACs require regular maintenance to prevent mould growth in the supply air duct and ensure the precise functioning of sensors. These practices are mandatory requirements for maintaining a healthy and comfortable indoor environment in buildings. The HVAC system defects are categorised according to issues concerning (a) the chiller plant (e.g. compressor problem, insufficient/slow cooling, chilled water pipe and condenser pipe leakage and condensation); (b) cooling towers (e.g. biological fouling and Legionella outbreak); (c) air handling unit (e.g. air distribution system efficiency and noisy operation/excessive vibration); (d) air distribution and terminal systems (e.g. dirty and mouldy ductwork and filter media choked at the air terminal). Discussion on accessibility issues concerning the HVAC system is added in a box section. As periodic system failure is expected in a HVAC system, the restoration of the system to its normal operable state in a given timetable is important. Disregarding preventive maintenance for components which require energy resources (e.g. chillers and fans) will reduce the HVAC's service life, resulting in untimely HVAC component replacement, and spark a surge in operational costs [3]. The maintainability of HVAC systems could be considered during the design stage to ensure adequate access for the inspection, maintenance and repair of all components.

The sanitary and plumbing systems' main functions are to provide hot and cold water and remove liquid waste or liquid borne solid waste [4]. For any water supply, especially potable water, any pollution, obnoxious taste or odour and toxicity should be completely prevented. In order to ensure the health and safety of the public and cleaning personnel, it is mandatory to avoid any blockage or flooding of sewerage system by controlling the surcharge [5]. The identified sanitary and plumbing systems' defects pertain to (a) general plumbing issues (e.g. water supply general defects and leaky joints in inaccessible area); (b) water supply system (e.g. corrosion and scaling of pipe/valve, damaged piping, water hammer and excessive vibration); and (c) water tank (e.g. accessibility, leakage from water tank, corroded water tank body and overflow of water). Design changes concerning sanitary and plumbing systems should not be encouraged since it is highly expensive and challenging to implement during the operation stage. Proper design, construction and maintenance of sanitary and plumbing systems should also put extra emphasis on ensuring a high standard of health of building users.

Fire protection systems are standardised elements as approved by the local fire department. For a performance oriented design of fire protection system, various recognised guidelines such as the Singapore Fire Safety Engineering Guidelines [6] can be consulted. The fire authority is notified in advance of any shutdown due to maintenance, especially when the automatic sprinklers are under maintenance. Such shutdowns could be scheduled during normal working hours and finished in the shortest possible time. If required, special precautions (e.g. fire petrol, temporary water connection, etc.) could be employed [7]. The fire safety defects raised in this chapter concern (a) fire detection issues (e.g. faulty fire alarm panel, faulty manual call point, faulty detector, inaudible/unidentifiable alarm, and backup power/lighting during power outage for fire protection); (b) faulty fire hydrant systems (e.g. damaged fire hose, accessibility problem of fire hydrant, and faulty fire hydrant point); (c) sprinkler system issues; and (d) faulty fire extinguishers (e.g. poor discharge of portable fire extinguisher and obstructed fire door). Considerations for the maintainability of fire protection systems at the design stage could prevent the occurrence of these identified defects. Making sure that the fire protection system is in proper condition could help avoid inconveniences during cases of false alarms.

The design, installation and maintenance of electrical systems differ significantly based on power rating. In the Code of Practice for Electrical Installations, 600V is considered low voltage and is usually employed in commercial buildings in Singapore [8]. The usual electrical system components are: transformer, wiring, distribution equipment, protective device, lighting, standby/emergency power supply, and grounding and lightning protection system [9]. Many defects of electrical system occur due to other services that share the same service spaces (e.g. seepage through building components, mismatch of lighting grid, etc.). Information regarding these critical aspects should be conveyed to the relevant professionals at the outset of the project. The identified electrical system defects refer to (a) switchgear problems (e.g. unsafe switchboard/electrical power distribution); (b) standby power generator issues; (c) artificial lighting and control system problems; (d) Lighting Protection System and Earthing defects. Integrating predictive maintenance and maintainability issues at the design stage could maximise the electrical systems' efficiency as well as the effectiveness of its performance throughout a building's life cycle.

An elevator is a machine that is installed rather than constructed and used for transporting people and things vertically to different levels in a building. Owing to its design requirements, traction elevators demand more maintenance time compared to all other vertical transports, while a hydraulic lift is 2–4 times simpler to maintain; however, its use is restricted to only eight floors [10]. There is no straight forward rule to planning for the most efficient elevator. The maintenance scope for elevators may involve lubrication, cleaning, checking, setting, adjustment, repair or changing or parts, and passenger rescue operations under normal circumstances. The identified defects pertain to (a) general issues (e.g. elevator machine room condition, elevator pit and lift lobbies); (b) common faults such as elevator car landing gaps and faulty door operation; (c) safety and reliability issues, faulty suspension ropes due to overloading, and failure to activate overspeed governor; (d) energy efficiency and lighting concerns; (e) general requirement for escalators and passenger conveyors. Considerations for maintainability when designing the vertical transport system could ensure its optimal required performance while simultaneously minimising costs for maintenance and operations throughout the building's life cycle.

6.1 HVAC System (ACMV plant and equipment)

All designs, construction, installation, testing and commissioning (T&C), operation and maintenance (O&M) of air conditioning and mechanical ventilation (ACMV) Systems must comply with the Code of Practice for ACMVs in Buildings, SS 553:2016. For the energy efficiency standard for building services and equipment, SS 530:2014 must be adopted.

6.1.1 *Chiller Plant*

Problem	Design	Construction	Maintenance
Chiller frequently unloading (stop-start); Compressor not starting  Rust stain  Leaking condenser pipes	Conform to the minimum requirements for air- conditioning equipment efficiency as per SS 530: 2014. (See also ANSI/ ASHRAE/IES Standard 90.1-2016). Specify proper placement for installation of temperature sensors/ probes to capture realistic average temperatures of spaces. Conform to the safety and environmental requirements for refrigeration systems (ISO 5149-1:2014/Amd 1:2015).	For applications with high dynamic load conditions of a facility, install and commission chillers with variable speed drive (VSD) compressors. Ensure that schematics and maintenance regime of refrigerant leak detection system are handed over to maintenance personnel upon completion. Set up appropriate cut-in and cut-out temperatures in chiller to avoid frequent unloading.	Inspect evaporator tubes for excessive oil, dirt or frost; check operating condition of expansion valve; check condenser tubes for air, dirt, scale, and sludge, and clean/ purge if necessary. Check condenser water supply and cooling tower efficiency. Inspect overload relay, and the condition of high-pressure and low-pressure cut-outs. Conduct monthly inspection on refrigerant level to avoid low pressure cut-outs of chiller (ANSI/ASHRAE/ ACCA Standard 180-2012).
Insufficient/slow cooling  Scale in condenser pipes	Design chiller plants according to the required cooling demand, with provisions for future expansions as per SS 553: 2016 and in compliance with the requirements of SS 554:2016. (See also AS 1668.2-2012, ISO 16814: 2008).	Conduct proper commissioning of chiller plant and set reasonable points. Perform post-installation monitoring of the installed instrument's performance through the BMS (building automation system) or EMS (energy measurement system).	Inspect temperature controllers and thermostatic control valve for any malfunctions and then reset. Perform routine inspection of condenser pipes and clean and conduct servicing (de-scaling) if/when necessary.

(Continued)

(*Continued*)

Problem	Design	Construction	Maintenance
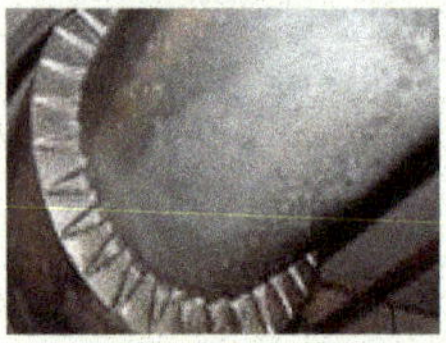 No insulation for chilled water pipe  Leakage of refrigerant	Comply with air conditioning plant optimisation strategies (e.g. high-efficiency chillers, aggressive condenser water reset, medium temperature chilled water loop, chilled water VFD pumping, etc.) as per SS 564-1:2013. Use refrigerants with zero ozone depleting potential (ODP) or global warming potential (GWP) of less than 100 [2]. Recommend installation of leak detection system in critical areas of plant rooms.	Document as-built drawings (including concealed services) for building user phase. Prepare and handover maintenance checklists for service and repair of each instrument during commissioning. Refer to measures for recommended monitoring procedures for chiller efficiency as per SS 591: 2013.	Routinely inspect insulation for any damaged/worn-out layers. Daily logging of chiller system to ensure system operates at optimal conditions. Any deviation from the intended chiller operation or alarms need to be attended to promptly. Perform annual shutdown or overhauling as per manufacture's guidelines (ANSI/ ASHRAE/ACCA Standard 180-2012).
Chilled water pipe and condenser pipe leakage and condensation 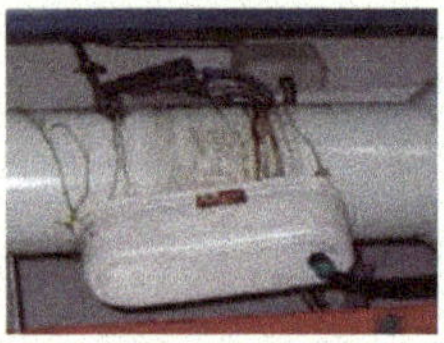 Leaking motor valve	Conform to the pump system's design's calculation and required outputs as per ISO 13612-1:2014. Specify the provision of flow meters for chilled/condenser water loops. Specify the use of unplasticised PVC pipes for cold water services/ industrial use as per SS 141:2013 (see also ISO 1452-1:2009).	Ensure proper workmanship during pipe installation and testing, especially at joints. Perform proper insulation of chilled water pipes to avoid condensation. Install primary-only variable flow chilled water pumping systems (SS 564-1:2013).	Conduct quarterly check for pitting noise in pumps. Avoid harsh cleaning methods that may damage pipes or cause the thinning of pipes. Prepare maintenance checklist for flow meters (ultrasonic and full-bore magnetic in-line types) (SS 591:2013). (See also ISO 12242:2012).

ACCESSIBILITY CONSIDERATIONS IN HVAC SYSTEM

Not enough clear headroom for access

Easy access to filters for maintenance

Storage of combustible materials in AHU room

Accessibility is a major concern in air-conditioning and mechanical ventilation design as these systems use a lot of plant space and equipment throughout various components of buildings, which drastically differ from one building use to another. Thus, considerations made for accessibility for equipment installation and subsequent inspection and maintenance are very important for a HVAC system's maintainability.

Access routes with safe access need to be rigorously planned for equipment maintenance. Adequate headroom (>2m) and designated access ways (e.g. stairs, catwalks, marked hatches) need to be provided for safe maintenance access as per Building Control Regulations.

Design should also accommodate requirements to minimise the spread of fire and smoke via the ductwork system. Ensure the duct covering, lining and flexible connection materials are non-combustible (SS 553:2016, AS 1668.2-2012).

During construction, ensure maintainable components of plant and equipment are installed in accessible positions. Also, check accessibility for inspection and recalibration of sensors during commissioning.

Provide appropriate working space (min. 600mm all round) for routine maintenance and equipment (including parts) replacement. Identify and demarcate any obstructions which result in reduced headroom or clear space around plant and equipment. Perform effective housekeeping and safe working conditions without congesting the maintenance spaces.

6.1.2 *Cooling Towers (CT)*

Problem	Design	Construction	Maintenance
Biological fouling of Cooling Tower 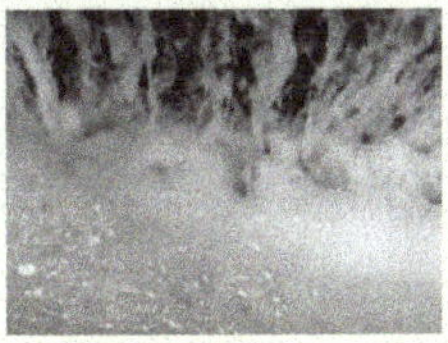 Foaming in cooling tower basin	Provide a suitable cooling tower design to prevent dirt accumulation and stagnation. Consider minimising tower fan power and size of towers for close approach as per SS 564-1:2013 in the design scheme.	Prepare and submit an Operation and Maintenance (O&M) manual for the chemical dosing system after the successful commissioning of the system. Install equipment to routinely observe	Perform frequent overall visual inspection and cooling tower sequencing. Clean tower fill, basin and drift eliminators. Conduct weekly check on fan motor; clean screen; make up water float and water sampling.

(*Continued*)

(*Continued*)

Problem	Design	Construction	Maintenance
 Scale and dirt build-up in fouled fills	Specify an automatic chemical dosing system that is capable of checking the conductivity and other required parameters of the CT daily. Ensure that the CT's layout has ample space for cleaning so as to prevent bacterial growth.	chemical tank levels and the tank condition of the automated dosing system. During installation, ensure that access hatches, level indicators, mixers, pumps, etc. can be easily reached by personnel for maintenance in the operation phase.	Conduct monthly check of motor supports, fan blades, motor alignment. Check on condition of bearings and motor, as well as for nozzle clogging, annually. Clean cooling tower at least once a year. Perform monthly check for legionnaire, scaling and corrosion in/of the condenser system. Disinfect and manually de-sludge cooling towers if required.
Legionella outbreak  Outbreak on air-water surfaces of Cooling Tower  Cooling tower air intake blocked by organic matter	Avoid redundant pipework, bends, and loops for cooling system design. Allow easy access to all parts of the system for maintenance [11]. Minimise drift by enclosing cooling tower pond. Take the wind direction/ distribution and the surrounding environment into account when designing the cooling system. Use non-corrosive, chemical-resistant, non-porous, smooth, opaque (to block sunlight) material to inhibit growth and proliferation of microorganisms.	Perform proper commissioning to ensure safe operation of cooling towers. Develop Cooling Tower maintenance manual including cleaning/water treatment/ decontamination procedures and handover during commissioning. Ensure system is clear of dirt/debris/organic matter and clean before operation.	Monitor cooling tower's water temperature, since elevated temperatures and moisture at air-water surfaces provide ideal conditions that may serve as a nutrient source for legionella growth. Check and conform with chemical concentration limits of cooling tower effluents as per local codes and regulations (e.g. Sewerage and Drainage (Trade Effluent) Regulations 1999). If cooling tower is not in use, it must be kept drained and dry. If not in use for more than 5 days, it should be drained, cleaned and

(*Continued*)

(*Continued*)

Problem	Design	Construction	Maintenance
 Testing the water for legionella	Design of cooling towers must be capable for preventing dirt accumulation and water stagnation. Accommodate suitable and efficient water treatment methods in the system's design.		disinfected before operating. Conduct regular testing for legionella bacteria, and get water samples from the cooling tower pond (ASHRAE 12-2000, ANSI/ASHRAE 188-2015.

6.1.3 *Air Handling Unit (AHU)*

Problem	Design	Construction	Maintenance
Air distribution system efficiency  Broken fan belt  Blue/grey formicary corrosion on Al-Cu cooling coil	The design of an AHU system must take into consideration energy efficiency and maintainability when selecting the system and its components. The design of the AHU must comply with the requirements of ANSI/ASHRAE/IES Standard 90.1-2016. (See also SS 553:2016). Fan power limitation requirements in mechanical ventilation systems as per SS 553:2016 must also be complied with. (See also AS 1668.2-2012). Specify minimum performance for air-conditioning equipment (including test procedure) as per SS 530:2014. (See also ANSI/ASHRAE/IES	The installation (and subsequent operation) of the AHU should comply with the fire safety requirements of the Code of Practice for Fire Precautions in Buildings (SS 553:2016). The vibration control of fan systems is to conform to the requirements of SS CP 99:2003 upon installation/commissioning. (See also ISO 9996:1996). Install/integrate AHU system components (e.g. VFD, control dampers/actuators) off-site to ensure quality installation. Merge continuous commissioning with the preventive maintenance	Clean dirt from impeller, fan scroll and blower blade, washable filters, filter frames and AHU frame slot. Clean cooling coil face with water. Flush with chemical cleaner, but avoid over dosage. Conduct chemical wash followed by thorough cleaning with water. Check for any dents on the coil fins. Comb/replace as needed. Clean/maintain fan and motor shaft for fire precaution [9]. AHU rooms should not be used for storage, and should avoid housing installations that are not associated with the air-conditioning system (SS 553:2016, AS 1668.2-2012). Inspect flanged joints for possible air leakages.

(*Continued*)

(Continued)

Problem	Design	Construction	Maintenance
	Standard 90.1-2016). Propose a biological film application on cooling coil.	programme to ensure optimal operation (ANSI/ASHRAE/IES/ USGBC Standard 189.1-2014).	During maintenance activities, clean the aluminium foil surfaces when applying ducting to ensure adequate bonding in order to conceal joints and avoid air leakage.
Noisy operation and excessive vibration  Noisy fan motor  Good practice: Install a connection box on AHU for vibration monitoring	Refer to the BS EN 1992-1-1:2004+A1:2014 for avoidance of discomfort or alarm to occupants, structural damage, or interference with proper function. Careful design is necessary to reduce vibration disturbances caused by impact force. Conform to noise control and elimination guidelines. Use principles of vibration control in determining noise control measures (SS CP 99:2003, AHRI 260, ISO 9996:1996).	Install acoustic silencers, shock isolators, machine enclosures and partition walls to reduce noise disturbances. Install vibration isolators/ mounting of equipment to minimise transmission of vibration from machine to load-bearing structure of the building (AHRI 260, SS CP 99:2003, ISO 9996:1996). Recommend installing vibration monitoring equipment for AHU to assist in predictive maintenance.	Conduct inspection to determine noise levels (AHRI 260). Inspect chiller plant with maintenance experts if abnormal sound occurs. Provide isolation/ mounting systems to reduce mechanical vibration and shock as per ISO 2017-1: 2005. Vibration from bearings, and sliding and rolling friction could be reduced by maintaining minimum clearances and proper lubrication (SS CP 99:2003, AHRI 260). Conduct regular monitoring of AHU fan vibration and compare with the vibration readings specified in ISO 10816-3:2009.

6.1.4 *Air Distribution and Terminal Systems*

Problem	Design	Construction	Maintenance
Dirty and mouldy ductwork  Mould growth in duct  Damper inside AHU collecting condensation, leading to biological growth	Conform to the design requirements for air duct systems, including for fittings and accessories, as per SS 553:2016 (where consideration is given to air velocities in ducts, materials and construction method, etc.). Fulfil the compliance requirements of thermal insulation for pipework. Comply with the ductwork seal requirements. Ensure that rigid ducts are manufactured from steel, aluminium, glass-fibre or mineral wool, or other approved materials. Conform to the minimum duct insulation R-values for cooling supply/return ducts as per SS 553:2016. (See also AS 1668.2-2012).	Comply with the construction and installation requirements for air duct systems and their fittings and accessories. Ensure that the ducts or duct linings (where glass fibre or mineral wool is exposed to the air stream) are suitably protected to prevent fibre erosion. The ducts should be sturdily supported — provide metal hangers and brackets for supporting ducts. Guarantee that the inner surfaces of the ducts for supply and return air are smooth and resistant to abrasion in order to reduce dust accumulation (SS 553: 2016, AS 1668.2-2012).	Conduct noise and air loss check when necessary. Perform testing and rating of performance of ducted air-conditioners (ventilation, exhaust and leakage air flow) as per ISO/FDIS 13253. Conduct yearly check of ducting insulation. Perform duct leakage test (for ducts designed to operate at static pressures in excess of 750 Pa) as per industry requirements (SS 553:2016, AS 1668.2-2012).
Filter media choked at the air terminal  Choked filter	Provide locations for intake and return air terminals. Conform to the minimum filter requirements with regard to the use of air filters for cleaning outdoor air (MERV) so that no unfiltered air can enter the air handling system.	Conduct alternate efficiency test method for filters (ISO 29463-5:2011). Perform a one-time vet airflow test during commissioning and keep proper record of the report to ensure the cooling capacity is	Monitor air pressure and pressure drop across the filter and replace filter when needed. If filter is heavily dented, the filter should be replaced. Conduct leakage test to test filter for local penetration ISO 29463-1:2011.

(*Continued*)

(*Continued*)

Problem	Design	Construction	Maintenance
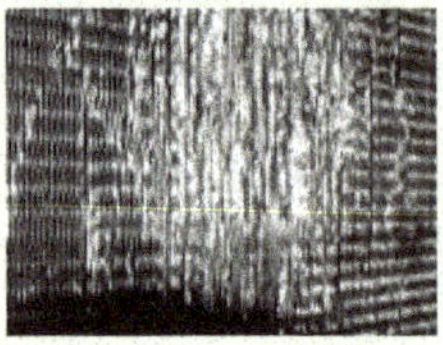 Bent aluminium fin	Follow the recommended use of a secondary filter of 80% dust spot efficiency (SS 553: 2016, AS 1668.2-2012). Provide sufficient access space for easy cleaning and replacement of filters.	adequate as per design load in accordance to the BCA Green Mark requirement, and that the noise level is within the desired level in accordance to NEA noise pollution standards [10].	Conduct monthly check and observe the dust accumulation level in ducts or grilles. Provide a display system for temperature and relative humidity at each floor and at each tenanted area.

6.2 Plumbing and Sanitary Systems

6.2.1 *General Plumbing*

Problem	Design	Construction	Maintenance
Water supply — general defects  Leaky joint wetting the floor  Ceiling damaged by pipe leakage due to haphazard pipe laying	Conform to the pipe sizing requirements based on hydraulic design and pump performance. Provide allowance for head loss, and frictional loss due to internal roughness, loss at fittings, turbidity, surge and pumping facility. Do not oversize piping as slow flow will cause stagnation. Specify standard fittings such as tees, elbows, etc. (BS 7291-1:2010, BS EN 598:2007+A1:2009, BS EN 545:2010).	Comply with the installation of water fittings as per PUB guidelines as stipulated in SS CP 48:2005. (See also BS 8558:2015). Fittings that are fabricated by welding together segmented pieces are not recommended. Avoid haphazard pipe laying.	Perform thorough investigation to check compliance with SS CP 48: 2005, and BS 8554:2015. Conduct monthly inspection of water flow rate and pressure, and position and functioning of valves. Conduct maintenance inspections of the pipe installation and identify/rectify physical defects such as broken pipe braces, dents, or leaks. (See also BS 8558:2015, BS EN 806-5:2012).

(*Continued*)

(*Continued*)

Problem	Design	Construction	Maintenance
Leaky joints in inaccessible areas 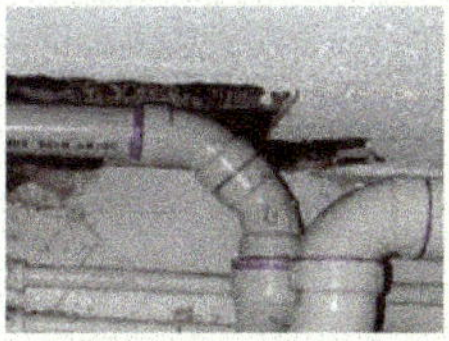 Leaking pipe above ceiling  Leaky water pipes in the ceiling	Adhere to the recommended requirement to maintain sufficient distance (> 400mm) from structure or other services running parallel to each other, for easy maintenance and to avoid interference or damage. Pipes should be of adequate strength and durability, and adequately supported (BS 8558:2015, SS CP 48:2005)	Ensure that the pipes and fittings are stored and installed as per manufacturer's instructions. Prevent any interior contamination. If contamination occurs, clean before installation. Take special care when joining two dissimilar materials.	Perform routine monitoring of piping system and check for any possible water seepage through the building fabric (BS 8558:2015, BS EN 806-5:2012). Repair the pipe joints properly using the correct jointing method. Tighten valve stems by replacing/fitting any missing gasket/washer.

6.2.2 *Water Supply System*

Problem	Design	Construction	Maintenance
Corrosion and scaling of pipe/valve 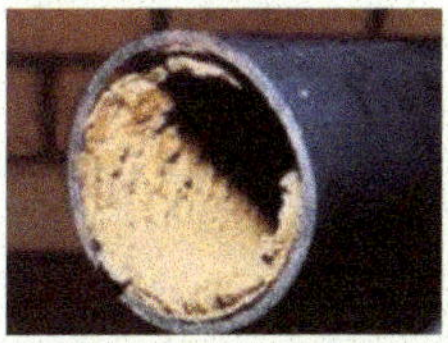 Scaling in pipe  Corrosion of pipe flange	Specify materials that are resistant to corrosion and non-reactive to the conveyed water and surrounding ground (BS EN 545:2010, BS EN 598:2007+A1:2009), and that do not impart any taste or toxicity to the water (BS EN 1796:2013, SS 375-1: 2015). Use of single material for the entire system is preferable for easy connection/jointing. Specify pipe system material that does not react with the pumped medium.	Pipe penetrations and joints should strictly comply with manufacturer's instructions. Jointing material should not enter pipe. Caulking at penetration sleeve should be made watertight. Fulfil proper installation while ensuring that the protective coating is not lost/damaged during installation (BS 8558:2015, SS CP 48:2005).	Conduct quarterly chemical and bacteriological analysis of water used [12]. Perform monthly check of water supply for visual signs of leakage, scaling and corrosion of pipes, joints and valves. Increase frequency of inspection for damp or polluted areas (BS 8554:2015). Conduct monthly inspections and clean off/remove any rust or scale. Re-paint parts in a timely manner if needed (BS EN 806-5: 2012).

(*Continued*)

(*Continued*)

Problem	Design	Construction	Maintenance
Damaged piping  Burst outdoor pipe  Damaged outdoor pipes  Corroded galvanised piping	Pipework design should consider factors such as the choice of material, rate of flow, accessibility, protection against damage, corrosion, avoidance of airlocks, water hammers, noise transmission, unsightly arrangements, vibration and expansion of fluid, stress and strains, etc. (BS EN 1057: 2006+A1:2010, BS EN ISO 21003-2: 2008+A1:2011, BS EN ISO 21003-3:2008, BS EN ISO 21003-5:2008). Provide adequate longitudinal support to pipe installations below ground to cater for loads and traffic vibration.	For underground pipe laying, bedding should be fully compacted prior to installation and the correct depth of trench, gradient, width and bottom condition should be maintained. Properly align pipe work and use suitable joints. Ensure careful backfilling at an adequate depth for underground pipe laying. Completed sections should be tested for defects using leakage tests and should be rectified by the contractor as required. Maintain proper water pressure in piping system to avoid bursting from over pressurisation.	Conduct thorough cleaning and disinfection of service pipes on a monthly basis, and clean the main pipes semi-annually. If required, removal of blockage with manual cleaning method (e.g. plunger, drain rod, spring auger) should be performed. Conduct chemical de-scaling quarterly (care should be taken so that it does not harm the pipes or jointing by giving consideration to chemical type or contact time). Perform maintenance schedule to check for clogged outlets (BS 8558:2015, SS CP 48:2005, see also [5]).
Water hammer  Cast iron pipe cracked by water hammer  Thrust bearing of a submersible pump ruined by water hammer	Pipework design should avoid water hammers (water pipe banging) (JIS A 4422:2011, BS EN 1057:2006+ A1:2010). Design airlocks and low supply pressure to minimise turbulent flow. Evaluate the required strength of valves and tightness of body and in between the inlet and outlet chamber as per BS EN 1567:1999.	A water hammer may arise when the electric valves on appliances or single control valves are shut off fast. Although all noises due to water flow and pipe expansion cannot be removed, the contractor is responsible for fastening the pipes properly and commissioning valves/ actuators to minimise the water hammer. Install check valve to control the creation of a vacuum in discharge pipe (BS EN 16767:2016, BS EN 545: 2010).	Conduct monthly pressure test by operating pump for min. 1 hr with 125m head or 150% of working pressure (whichever is greater) and check for any individual leakage or overall leakage. Conduct semi-annual pressure test for sewage pumps. To rectify water hammer issues, provide additional bracing or anchor block support at bends and branches in order to withstand the hydraulic thrust.

(*Continued*)

(Continued)

Problem	Design	Construction	Maintenance
Excessive vibration  Insufficient vibration damping causing wear-out of parts  Dented pump impeller	Ensure that pumps are properly sized to meet the required pressure. Head loss, frictional loss and loss at bends should also be considered. Sewage pumps should be able to handle long and fibrous material. If required, pre-treatment (e.g. crushing) is provided. Specify variable speed drives (VSD) that are able to handle both maximum and variable demand. Ensure that the pump is mounted on an isolation bed (e.g. 150mm insulated padding) and that no site adjustment in height or position should be done.	Conform to the installation of pumps as per manufacturer's instructions. Ensure proper construction of isolation beds and installation/mounting of equipment. Special attention should be given for tightness of joints, and alignment of bearings and pipes. Conduct testing and commissioning of the auto and manual interchange of duty and standby pumps.	Lubricate pump parts with oil or grease as per manufacturer's instructions. Conduct routine check, clean motor starter and all heavy current contacts, and replace worn parts. Visually check for any damage or missing parts (screw, nuts, strainer, etc.). Ensure that after installation, pump is tested with power on mode for any unusual vibration, noise, leakage or burnt smell.

6.2.3 *Water Tank*

Problem	Design	Construction	Maintenance
Unauthorised/poor accessibility  Lack of safety cover and security lock in water tank	Provide a permanent climbing ladder for easy inspection and cleaning of interior. Provide a minimum of 600mm space, on all sides of the tank. Install minimum possible number of openings to each compartment; each opening should be fitted with a cover/trapdoor.	Ensure there is adequate access space to and around the tank. Ensure that the tanks are not compartmentalised, so as to avoid the shutting off of the whole supply during cleaning (inconveniencing building users).	Conduct general housekeeping within and around the tank room to remove any obstructions to maintenance access. Perform monthly cleaning of wash-out pipes to ensure proper flushing out of the water.

(Continued)

(*Continued*)

Problem	Design	Construction	Maintenance
 Good practice; make provisions for access to water tank(s) for maintenance and/or inspection	Provide corrosion-resistant, mosquito-proof netting for overflow pipe/vent. Design a series of tanks instead of one large tank, to meet demand for isolation during maintenance, for ease of access.	For the purpose of safety and security, provide a lock to water tank, so that only authorised personnel can access it. Ensure that the access is easy and safe, and for authorised persons only.	Remove sand and dirt deposits in cisterns and tanks. Remove rust stains and repaint affected parts as required. Prevent dirt, dust, insects, birds, etc. from entering the tank. Conduct thorough cleaning and disinfection of tank interiors semi-annually.
Leakage from water tank  Corrosion of bolts and subsequent leaking of GRP tank	Ensure the proper design detailing of pipe penetrations at the tank to avoid leakage. Avoid cracks in concrete tanks; ensure water-tightness, through proper structural design. Specify appropriate waterstops and sealants where pipes penetrate the structure. Maintain a minimum air gap above maximum water level (SS 245:2014, BS EN 13121-3:2016).	Construct the tank body as per specifications (additive, coating, and lining), and render as a monolithic and watertight container. Maintain the exact size and positions of installed devices (SS 245:2014, BS EN 13121-3:2016). Commission tank by testing for water-tightness; check for any leakage, seepage, and water loss. Ensure that all components are functioning well.	Inspect drainage lines and basin. Conduct routine checks for rusting of metal tanks and apply anti-corrosive paint/coatings where necessary. Perform routine checks on float valve and liquid level indicator for damages to avoid potential overflow.
Corroded water tank body  Corroded tank body	The body of the water tank should be made of watertight and corrosion-resistant material, such as reinforced or pre-stressed concrete, steel, and glass fibre reinforced plastics (SS 245:2014, BS EN 13121-3:2016).	Structure of water tank should be constructed with adequate strength and be free from any deformation. Refer to BS EN 10088-2:2014 for standards for a Stainless Steel Sectional Water Storage Tank (Minimum Grade 316).	Ensure that the parts (e.g. pipes/strainer) are corrosion-resistant and can be replaced over time. Perform timely re-application of coatings to avoid peeling and delamination of coat

(*Continued*)

(Continued)

Problem	Design	Construction	Maintenance
 Corroded and damaged water tank  Corroded water tank	Specify surface treatments, waterproof coating, or lining coating to resist water seepage and weathering. Ensure that such finishes do not affect the stored water's quality of hygiene (BS 8558:2015, SS CP 48:2005).	The water storage tank's installation must be certified by a Professional Engineer to ensure that it is structurally sound with regard to hydrostatic pressures, deflection and leakage.	(for steel tanks), and water seepage (for concrete tanks). Use disinfectant to clean water tanks (BS EN 805:2000). Once disinfectant has been sprayed on inner surfaces and pipes for the designated period, it should be thoroughly cleaned/removed.
Overflow of water 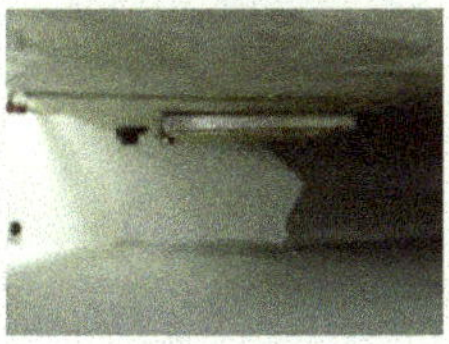 Overflow of water as seen from underneath a tank	Decision on the size(s) of the tanks should be made based on water demand, supply, probability of pump failure, time needed for repairs, ratio of peak hours to average flow rate, provision of alternative supply or storage, etc. (SS CP 48:2005, BS 8554:2015).	Ensure that the tank is capable of handling various loads (as applicable) without showing cracks, stress or deformation (SS 245:2014, BS EN 13121-3:2016).	Conduct monthly inspections to check the operation of float valve or any other effective device for controlling the inflow of water. All valves should be periodically operated to ensure that the working parts are moving freely (BS 8558:2015, SS CP 48:2005). Inspect condition of overflow warning alarm for the water tank. Inspect the condition of warning alarm which indicates when water levels fall below 50mm from the invert level of the pipes.

6.3 Fire Safety

6.3.1 *Fire Detection*

Problem	Design	Construction	Maintenance
Faulty fire alarm panel 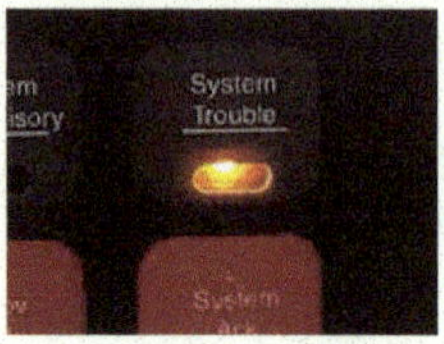 Fire alarm system fault indicated in fire panel. A faulty fire panel can cause fire incidents to go unmonitored 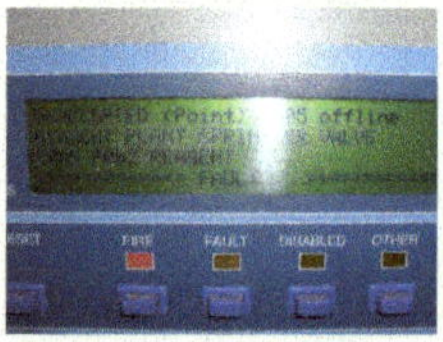 Control panel displaying fault message and indicating that detectors are offline	Design system to accommodate false alarm management as per SS CP 10:2005 and Fire Code [8]. Locate fire alarm panel in corrosion resistant cabinet without any exposure to excessive dampness. Fire panel connectivity should be independent and compatible with the building automation system (BAS). Specify red wiring for fire alarm system. Segregate from other ELV cables to remove electromagnetic interference. Use alarm verification features to reduce incidence of false alarms.	Installer should identify circumstances that can lead to a high rate of false alarms and should inform both the designer and user. Check to ensure acceptable levels of false alarms during commissioning (SS CP 10:2005, BS 5839-1:2017, BS 5839-6:2013). Install neatly and protect with sleeve as per manufacturer's requirements (care to be taken in concealed spaces). For the protection of joints in junction box, refer to minimum joint requirements.	Conduct regular testing and inspection of fire alarm panel as per SS CP 10:2005, BS 5839-1:2017, BS 5839-6:2013. Conduct daily check to ensure normal operation, and to record and rectify any faults. Perform weekly tests to check battery and voltage conditions. Conduct monthly simulation of zonal fire and fault conditions. Clean fire alarm panel for proper operation and visibility (NFPA 72:2016). Keep fire panels safe and secure from unauthorised tampering.
Faulty manual call point  Dirty manual call point at an institutional building  Damaged manual call point	Designate locations of manual call points along all exit routes and at final exits (BS 5839-1:2017, BS 5839-6:2013). Manual call points should be located in such a way that a person should not have to travel more than 45m along an escape route to reach a manual call point, when the layout of the building is known. Each manual call point should be positioned 1.4m (+/-200mm) from floor level (BS 5839-1:2017).	Use pull station type manual call points in outdoor areas rather than break glass type, in order to prevent water ingress. Ensure that manual call points are securely mounted and properly aligned (SS CP 10:2005). Upon installation, test the system (e.g. three-second response test for manual call point, battery removal test, etc.) (BS 5839-1:2017, BS 5839-6:2013).	Conduct monthly test of manual call points on all alarm zones to ensure each part is functional, and especially check whether the remote auxiliary facilities are initiated or not. Monitor power supply and faulty wiring of call points and other elements of the fire detection system.

(*Continued*)

(Continued)

Problem	Design	Construction	Maintenance
Faulty fire detector 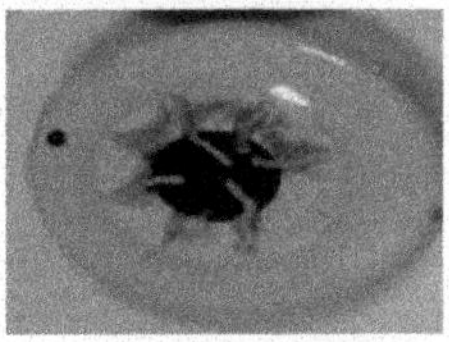 Dirty and damaged fire detector  Faulty, corroded and improperly installed detector	Comply with the fire detection system requirements of Fire Code 2013 by SCDF [8], SS CP 10:2005, ISO 7240 series, BS 5839-1: 2017, BS 5839-6:2013. Ensure detectors are accessible for maintenance and replacement. Conform to the selection of heat, smoke, and flame types based on requirements on-location. For decisions regarding the number of fire detectors, and their location and spacing design requirements, refer to SS 575:2012, BS 9990:2015.	Ensure the proper fitting of each fire detector, to avoid misalignment or damage caused by shock. Ensure quality workmanship so as to avoid detector obstruction and or detector being covered by paint. Remove paint, dust or any foreign material that can affect its function from detector.	Practice proper housekeeping to maintain cleanliness and avoid obstructions (especially at poorly accessible points). Inspect fire detectors weekly, and conduct monthly fire alarm simulations from a randomly selected detector to check the entire system. Perform annual test of 20% of all detectors; all detectors will be inspected over a five-year period (SS CP 10:2005, BS 5839-1:2017, BS 5839-6:2013).
Inaudible/unidentifiable alarm[1]  A site damaged by fire; fire propagated due to alarm system's failure to signal in time	Design fire alarm system (e.g. location, type and number of alarms) as per SS CP 10:2005 (see also BS 5839-1:2017); and emergency voice communication system as per SS 546:2009 (see also BS 5839-9:2011).	Establish, implement and maintain procedures for warning and communication (e.g. life safety); and set incident communication procedures (SS ISO 22313:2013). Link lifts with audible warnings and emergency detection system as per ISO/TR 25743:2010.	Inspect alarms for defects (e.g. loose or blocked gong bolt, damaged or corroded alarm, alarm spoilt by temperature fluctuations, etc.). Conduct annual check of all installed speakers, amplifiers, and connecting appliances (including cables) and keep records.

(Continued)

[1]The installed alarm system should emit a minimum sound level of 65dB(A) or 5dB(A) above background noise (if lasting more than 30 secs) and at a frequency between 500Hz and 1000Hz. The maximum sound level should not exceed 120dB(A). If the sound needs be carried through a door it should be taken into consideration that the decibel loss occurring through doors is approximately -20dB(A) through a normal door, and approximately -30dB(A) through a fire door (BS 5839-1:2017).

(*Continued*)

Problem	Design	Construction	Maintenance
	Alarm sound selected should be distinguishable from general clutter. Use a visual alarm where there is excessive background noise. Propose incident communication facilities (as per SS ISO 22313:2013) and determine internal/external communication needs (e.g. through PA system integrated with iBMS) as per SS ISO 22301:2012.	Installation and cabling of fire alarm devices should be done in such a manner where, in the event of a fault, at least one sounder located within the vicinity of the control and indicating panel will remain in operation (BS 5839-1:2017).	Conduct routine check of fire alarm panel indicator bulb operation and battery (NFPA 72:2016). Perform real time monitoring and management of system performance.
Malfunctioning or damaged backup power/lighting  Damaged exit signage; may interfere with egress during evacuation  Good practice: Regularly test emergency exit signs to ensure that they are in working order	Design emergency power supply systems in buildings as per SS 563-2:2010; and emergency lighting fixtures as per SS 563-1:2010 (see also BS 5266-1:2016, BS EN 50172:2004, BS 5266-8:2004, NFPA 101, AS 2293 SET:2005). Building emergency generator supply should be able to back up the emergency voice communication system (SS 546:2009, BS 5839-9:2011). A standby generator may be used solely to provide power to emergency lighting systems, or in addition, to meet requirements other than those directly associated with emergency lighting.	Conform to the requirements for the installation of emergency power supply systems in buildings as per SS 563-2:2010, BS EN 1838:2013, BS 5266-1:2016. Individual luminaires should be mounted to avoid glare and if possible, should be positioned at least 2m above floor level (measured from floor to the underside of the luminaires). The horizontal illuminance on the centre line of any exit cannot be less than 0.5 lux. A fuel supply must be readily available to ensure that emergency lighting operates continuously for the rated period following the failure of normal power supply (SS 563-2: 2010).	Ensure the maintenance of emergency power supply systems in buildings. Conduct a monthly manual test of emergency lights and replace batteries or lamps as soon as a fault is detected (SS 563-2:2010, AS/NZS 2293.2:1995/AMDT 3:2012). Conduct monthly fire simulation test. Simulate failure of main power supply and test the efficiency of the standby battery. Ensure visual and audible fault signals are activated once the battery is disconnected.

6.3.2 *Fire Hydrant System*

Problem	Design	Construction	Maintenance
Fire hose damaged (cut kink, leak, missing part, abrasion) 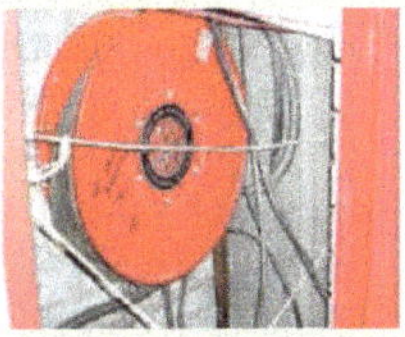 Haphazard winding  Damaged hose reel cabinet  Damaged hose reel  Damaged fire hose coupling	Ensure that the fire hose reel is suitable for the particular use of the facility in question, and that it complies with the related standards (BS EN 694:2014, BS EN 1947:2014). Comply with the technical quality acceptance for fire hoses as per BS 6391:2009. Fulfil cabinet specification (size and mounting) as per standard guidelines. Follow the distribution and number of fire hose reel cabinets as per SS 575:2012. (See also BS 9990:2015, NFPA 14:2016). Ensure that access to or visibility of fire hose is not obstructed. Fire hose cabinet should be made of maintenance-free fire-proof material. The location should allow for 180° opening of cabinet door. The wall mounted type is only allowed in riser main shaft.	Protect hose reels from mechanical damages. The reel should be mounted overhead, but the nozzle retainer, hose guide and inlet valve must be kept at 900mm above finished floor level (BS 9990:2015, SS 575:2012, NFPA 14:2016). During commissioning, the hose should be flushed out to remove harmful matter. Conduct flush out test to remove any kink or knot and to ensure that all valves and nozzles are operational. Ensure that reel brackets are firmly fixed, so that the hose can be used properly.	Perform proper housekeeping and avoid mishandling. Once a month, check for corrosion/leakage (of drum), and ensure that hose, nozzle, stopcock, hinges, break glass device and cabinet are in acceptable working condition. Lubricate as required. Conduct monthly water flow pressure test and annual hydrostatic test (BS EN 1402:2009) to check for defects or leaks, especially if the hose has been exposed to chemical or severe stress. During the test, the hose is completely run out and subjected to operational pressure. After the test, it should be dried and properly secured with a Velcro strap. Ensure that fire hose is stored in a cool, dry place (ISO 2230:2002).

(*Continued*)

(Continued)

Problem	Design	Construction	Maintenance
Accessibility problems (difficulty in accessing fire hydrant)  Inaccessible hydrant points due to positioning  Improper housekeeping practices, and storing of combustible material in wet riser	Conform to the location and number of fire hydrants as per Fire Code [8]. Comply with the positioning of breeching inlets as close as possible to rising main/hydrant (BS 9990:2015, SS 575:2012, NFPA 14:2016). Locations should be accessible, with no obstructions from parking, loading bays, landscaping, building elements, etc. Provide protection to hydrants from mechanical damage. Specify easily visible/ identifiable signage and colour as per SS 508-3:2013 (see also ISO 3864-1:2011).	Ensure good project coordination. Installation should be secure and safe, with special consideration given for potential sources of damage. Ensure connections and position of valves comply with specifications. Risers must be securely anchored before any pressure or flow test is performed. Hydrants should be made operable immediately after completion, and should be tested to protect the construction site. Mounting height of hydrant and breeching inlet should be strictly maintained during installation (BS 9990:2015, SS 575:2012, NFPA 14:2016).	Perform proper daily housekeeping practices at hydrant points to remove obstructions (debris, stacked material) that impede accessibility. Ensure that the storage tanks are accessible for maintenance. Ensure that the valve pit is accessible for inspection and cleaning. Conduct semi-annual check for rust, dirt, or foreign material on valves, or other operating parts; as well as clean, paint and lubricate as required. Ensure that additional building elements, landscaping, etc. (during building operation and maintenance phase) do not impede accessibility to hydrant points (BS 9990:2015)
Faulty fire hydrant point (damaged, jammed, leaky)  Damaged hydrant	Specify hydrant pillar to be constructed of materials that are strong and rust-proof (e.g. gunmetal parts) (BS EN 1982:2008).	Ensure the proper installation of all components (parts-stem, cap, plug, thread, etc.) without damaging them.	Conduct weekly check of isolating valves to ensure that they are kept locked in open position daily and that breeching inlets are functioning (NFPA 25:2017).

(Continued)

(Continued)

Problem	Design	Construction	Maintenance
 Damaged and unusable breaching valves	Rising main and other pipework should be made of wrought iron or steel. Pit covers on roadways should be able to withstand vehicular load (BS 9990:2015, SS 575:2012, NFPA 14:2016).	Lubricate and paint for additional protection. Tighten outlet properly after commissioning and testing.	Conduct monthly checks for any leakage, blockage or corrosion, and for workable line pressure. Perform thorough inspection of booster pump and associated systems semi-annually. Ensure that a thorough inspection of the hydrants is annually performed by a competent professional.

6.3.3 *Sprinkler System*

Problem	Design	Construction	Maintenance
Faulty/compromised sprinkler system  Leaky sprinkler pipe  Obstructed sprinkler	Sprinkler design requirements should consider hydraulic principles and parameters such as, hazard class, discharge density, and AMAO (assumed maximum area of operation) (SS CP 52:2004, ISO 6182 series, NFPA 13:2016). The usual requirement is 75L/min for 2.5m wide area. Specify rust resistant material to avoid corrosion, pitting and scaling. Potential obstructions should be considered during planning stage.	Installation and testing of sprinkler system, its associated controls, fire pumps and water supply should comply with SS CP 52:2004. (See also NFPA 13:2016). Mounting should be carried out according to the design approved by the authority and as per manufacturer's instructions. Conform to general guidelines as per NFPA 13:2016. Ensure the careful installation of sprinkler system to maintain correct orientation without hindrance by supports.	Conduct quarterly visual inspection of all sprinklers for any leakage, damages or grease/dirt in spray nozzle and replace as necessary. Conduct annual inspection of pipes and hangers for corrosion and mechanical damage (clean, paint or replace as necessary). Clean quarterly and remove any obstruction affecting efficient discharge from sprinklers. Check for any sign of corrosion or deposit of dirt, paint or foreign material (NFPA 13:2016).

(Continued)

(*Continued*)

Problem	Design	Construction	Maintenance
 Corroded and faulty sprinkler		Maintain spare sprinklers and sprinkler spanner after installation for future needs.	Practice good housekeeping to avoid stacking of material leading to obstruction of sprinklers. Conduct monthly test of smoke detectors and regular maintenance to ensure continuous operation

6.3.4 *Fire Extinguishers*

Problem	Design	Construction	Maintenance
Poor discharge of portable fire extinguisher  Low pressure indicated in the fire extinguisher. This often leads to poor or no discharge  Extinguishing cone missing  Fire extinguishers not kept properly and not secured	Ensure that the locations and number of portable fire extinguishers are based on the maximum travel route (SS EN 3 Series, BS 5306-8:2012, ISO/PRF 7165). Access to or visibility of extinguishers should be unobstructed. Extinguishers should be visible along an escape route (preferably near room exits, along corridors and staircases, in lobbies, and on landing). The extinguishers' body and the parts should be of approved quality to prevent rusting, early damage or deterioration (BS EN 3 series).	Ensure the proper positioning (designated location, hung properly with label facing out) of fire extinguishers. Installation of Small Fire Extinguishers (≤ 4 Kg): Hung on wall with hanger or bracket such that the handle is about 1.5m from floor. Hangers should be securely fixed. Installation of Heavier Fire Extinguishers (≥ 4 Kg): Carrying handle should be about 1m from floor. Ensure that the arrangement would not hurt the person carrying it. Parts should be attached as per manufacturer's instruction.	Conduct regular servicing to confirm required working condition. (Refer to the recommended schedules for maintenance as per BS 5306-3:2009). Conduct monthly inspection to ensure that the pressure gauge is in operative range and check for any sign of corrosion of the body of the extinguisher. Comply with the recharging frequency as per type of extinguisher. Charging, testing and maintenance of fire extinguishers must conform to SS 578:2012 specifications. Extinguishers must be recharged with the same agent only. No mixing or cross contamination allowed and no overfilling. (See also BS 5306-9:2015, NFPA 10:2013).

(*Continued*)

(*Continued*)

Problem	Design	Construction	Maintenance
Fire door obstructed  Exit door obstructed by stacking of material  Obstructed corridor  Wedged fire door	Fire door should be selected in terms of stability, integrity and insulation as per requirement of SS 332: 2007. (See also NFPA 80). Except for the fire door's thickness, internal construction, facing, edging, and construction technique, any other aspects of the fire door may be customised.	Fix fire door as per manufacturer recommendation. Fire door must be the same make and model as the tested prototype. Door frames installed during wall construction should be thoroughly grouted in cavity as deeply as possible with corrosion-proof anchor. Screws for attachment should be driven properly, and not hammered or placed in other positions (SS 332:2007, NPFA 80). Proper workmanship to avoid damaging/ jamming/sagging door (e.g. tilted hinge).	Inspect fire doors at least once a year, to ensure that self-closing mechanism functions as intended at all times. Check for and remove any door stoppers, or materials stacked near or by the fire exit door. Remove any obstructions. Check integrity of door leaf for superficial damage, structural damage and excessive bowing or deformation. Inspect hinges, latches, bolts and pull handle weekly. Automatic release mechanisms should be tested in accordance with BS 5839-3:1988. Ensure that egress is unobstructed in case of emergency evacuation as per NFPA 101.

6.4 Electrical Systems

6.4.1 *Switchgear*

Problem	Design	Construction	Maintenance
Unsafe switchboard/ electrical power distribution  Water seepage in electrical closet 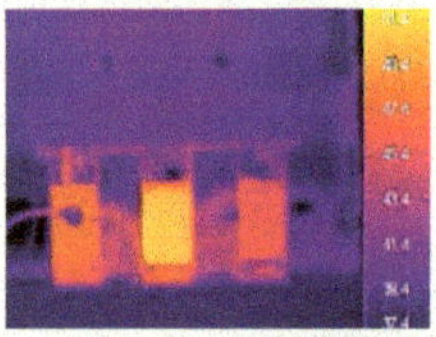 Thermal image of an overheated circuit breaker 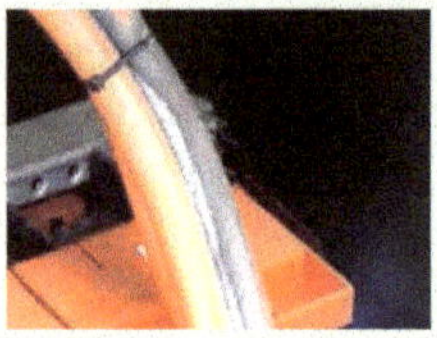 Burnt cable	Comply with the requirements of electrical installations as per BS 7671:2008+A3:2015, SS CP 5:1998, NFPA 70 [5]; including the location and number of power points. Ensure switchboards have adequate space and access for operation and maintenance. Specify suitable switch closets with regard to moisture exposure conditions. Refer to the definition of types and functionality of RCCBs and specifications for RCBOs as per SS 480:2016 (IEC 61009-1:2010+AMD1:2012+AMD2:2013 CSV BS EN 61009-2-1:1995). Provide sub-metering system with remote measurement capability and link to BMS/EMS to track energy consumption data trends [6].	Conform to the guidelines for construction and compliance inspection of electrical connections and earthing thereof (BS 1363-4:2016, SS 403:2013). Refer to BS 8512:2008 for storage, handling and installation of power cables on wooden drums. Install sub-metering system with remote measurement capability and link to BMS/EMS to track energy consumption data trends [6]. All accessible metal parts of connection units should be in electrical contact with the earthing terminal(s) (BS 1363-4:2016, SS 403:2013).	Conform to the maintenance of electrical installations as per BS 6423:2014, BS 6626:2010, SS 538: 2008, [3;7]. Check for insulation damages (e.g. cracks, blisters, warping) caused by overheating, physical impact or by spillage of cleaning chemicals. Check for potential short circuits or ground faults. Ensure that switchboards are not exposed to direct sunlight or alternative heat sources Conduct annual shutdown to eradicate hot spots along the distribution network as witnessed by the owner and certified by a Licensed Electrical Worker (LEW). Provide necessary warning notices/labels at switchboards (e.g. shock hazard warnings).

6.4.2 *Standby Generator*

Problem	Design	Construction	Maintenance
Standby power generator issues 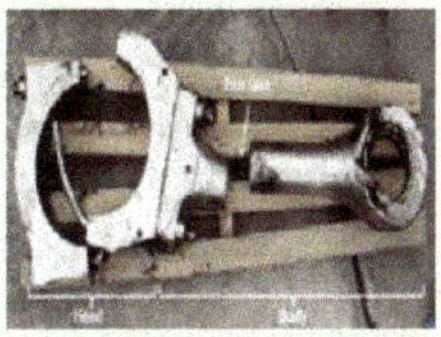 Failed connecting rod of diesel generator (Photo credit: Juarez *et al.*, 2016)  Burnt fuse  Storage of combustible material in generator house  Charging batteries of generator	Design mains failure standby power generation system as per code requirements. Provide sufficient headroom (>2600mm) in generator rooms for maintenance tasks — i.e. sufficient height to enable any portion of the generating set or equipment to be raised freely for dismantling — as per SS 535:2007. (See also BS 7698-7:1996, ISO 8528-7:1994, NFPA 110:2016). Comply with the general guidelines for earthing of generator sets and substations (BS 7430: 2011+A1:2015, SS 551:2009). Adhere to the recommendations for daily diesel service tank package and tank storage (i.e. safety and suitability of design, emergency provisions, and minimisation of vapour hazard) as per SS 532:2016. (See also BS 5908-1:2012).	Conform to requirements for the construction, installation and testing of generator systems for buildings. Provide sufficient access and clear passage for construction and maintenance. Allotted spaces must be of sufficient strength or suitably strutted to support the loads (SS 535:2007, BS 7698-7:1996, ISO 8528-7:1994, NFPA 110:2016). Install indoor fuel tank with a level indicator that can be easily accessed for observation. List and mark electrical wiring and equipment located near/within hazardous zones (i.e. day tank) (as defined by NFPA 70B [7] or IEC 60079) for installation in an appropriate manner (SS 532:2016, BS 5908-1:2012).	Practice proper housekeeping and avoid stacking and storing of combustible materials in the generator house. Maintain records of preventive maintenance activities in a secure manner. Conduct general inspections daily and check on fuel, lubrication and cooling systems. Perform a monthly running of the generator on no load for half an hour. Check battery charger, starting batteries and drive belt tension. Adhere to the requirements for operation and maintenance of standby generator systems for buildings. Once a year, run load test on the generator and check to ensure that emergency supply can support all essential emergency services (SS 535:2007, BS 7698-7:1996, ISO 8528-7:1994, NFPA 110:2016).

6.4.3 *Artificial Lighting*

Problem	Design	Construction	Maintenance
Faulty/compromised artificial lighting and control system  Discoloured outdoor lighting cover  Corroded metal conduit	Ensure that lighting design will improve energy and sustainability objectives of the building. Specify a centralised lighting control system that allows easy monitoring; or automate the system with a proper control strategy. Ensure that: — the lighting control is readily accessible. — lamp efficacies and ballast energy performance should meet the latest Minimal Energy Performance Standard (MEPS). — lighting power density is calculated for the building and, that it meets the lighting power budget in SS 530:2014 (see also ANSI/ASHRAE/IES Standard 90.1-2016). — Calculate daylighting as per BS ISO 10916:2014	Conform to the recommended illumination levels for office areas and task activities as per SS 514:2016 (see also CSA Z412-2000 (R2016), ISO 8995-1:2002/Cor 1:2005). Comply with specifications for luminaires, for general requirements and tests (SS IEC 60598-1:2016). Maintained illuminance depends on the maintenance characteristics of the lamp, the luminaire, the environment and maintenance programme (ISO 8995-1:2002/Cor 1:2005, SS 531-1:2006 (2013)). Display and ornamental lighting should be separately controlled.	Conduct routine checking of adequate lighting levels and maintain adequate lux for appropriate activities in accordance to users' needs and statutory guidelines (CSA Z412-2000 (R2016), ISO 8995-1:2002/Cor 1:2005, SS 531-1:2006 (2013), SS 514:2016). Practice proper housekeeping by dusting off and cleaning lamp surfaces. Conduct regular inspection of light fittings and replace if burnt-out. Consider group re-lamping if lamps in the same batch are failing. Conduct routine check on transformers and drivers of luminaires. Check exterior lights for corrosion, torn cables, compromised watertight seals and discolouration; take remedial action where needed.

6.4.4 *Lightning Protection System (LPS) and Earthing*

Problem	Design	Construction	Maintenance
Lightning protection system (LPS) defects (Corroded/exposed parts)  Unsecured clamp/joint  Corroded conductor	Adhere to protection measures to reduce risk of damage by lightning (e.g. injury to living beings, physical damage and failure of electrical and electronic systems) as per SS 555-1:2010, BS EN 62305-1:2011, NFPA 780. Provide external lightning protection system (LPS) to intercept direct lightning flashes to the structure. Conform to design considerations for system earthing, including selection of type of earthing system to be used. Material selection and minimum dimensions (for earth-electrodes to resist corrosion) must comply with SS 551:2009 (see also BS 7430:2011+ A1:2015). Provide lightning electromagnetic impulse protection measures (LEMP) as per SS 555-4:2010. (See also IEC 62305-4:2010).	Construct and install air-termination system and LPS as per IEC 62305-3:2010, SS 555-3:2010. Installation to be performed by certified LPS installers. Choose electrode locations that avoid the drainage of fertiliser and other materials into the area. Top soil should not be mixed with the backfill around an electrode. To avoid hazards to adjacent ground systems, the electrode system should either be of compatible metals or protected by adopting cathodic protection (BS 7430:2011+A1:2015, SS 551:2009).	Perform a thorough check on surge arrestors and the earthing system once a year, together with the annual shutdown. Monthly inspection must be conducted by a Licensed Electrical Worker (LEW). Such inspection should cover internal LPS to avoid occurrence of dangerous sparking within the structure caused by lightning current flowing in the external LPS or other conductive parts of structure (IEC 62305-3:2010, SS 555-3:2010).

6.5 Elevators, Escalators and Moving Walkways

6.5.1 *General*

Problem	Design	Construction	Maintenance
Compromised/poor condition of elevator machine room  Leaking ceiling of the machine room 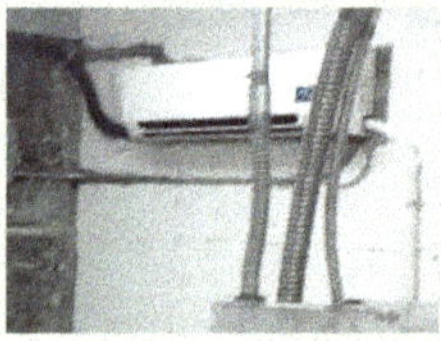 Good practice: Installing a cooling system in the machine room to prevent excessive heat in equipment	Comply with local codes, and consider elevator system performance (SS 550:2009, CIBSE Guide D [14], BS 5655-6:2011, BS 5655-11: 2005, BS EN 81-20: 2014). Provide ease of access to the elevator machine room with outward opening door (minimum clear opening of 0.6 × 1.8m) and permanent safe access for personnel and heavy equipment. Machine room should be equipped with electric lighting with a minimum illuminance of 200 Lux at floor level. Also provide switched socket-outlets. (BS 5655-6:2011, BS 5655-11:2005, BS EN 81-20: 2014, SS 550:2009). Provide good ventilation to the machine room (natural or mechanical). For natural ventilation, a 20% opening of the floor area is recommended to achieve cross-flow.	Provision for mechanical ventilation is recommended when the ambient temperature of the room exceeds 32°C. (BS 5655-6:2011, BS 5655-11:2005, BS EN 81-20:2014, SS 550:2009). Properly commission elevator prior to operation as per BS 8486-1:2007+A1:2011. Lift machine and drive must be securely mounted. All movable parts, the gear box, and joints should be sufficiently lubricated [13].	Conduct regular inspection of room condition and practice proper housekeeping. The room should not be used as storage; remove all non-elevator related materials from the machine room. Adequate lighting should be provided in the elevator machine room to allow workers to conduct maintenance works safely and efficiently [14]. Machine room should be ventilated to ensure the temperature difference measured at any point within 1000mm of machinery and associated equipment does not exceed 38°C (BS 5655-6:2011, BS 5655-11:2005, BS 7255:2012, BS EN 13015: 2001+A1:2008, SS 550:2009).

(*Continued*)

(*Continued*)

Problem	Design	Construction	Maintenance
Poorly-maintained elevator pit  Leaking elevator pit	Ensure proper waterproofing design of elevator pit (BS 5655-6:2011, BS 5655-11:2005, BS EN 81-20:2014, SS 550:2009). Specify corrosion resistant material and components in elevator system to minimise damage by presence of water or excessive moisture.	Test waterproofing of elevator pit before installation of elevator equipment (BS 5655-6:2011, BS 5655-11:2005, BS EN 81-20:2014, SS 550:2009). Avoid any damage to waterproofing membrane during elevator installation.	The pit areas should always remain dry. If there is any presence of water, the source of water must be identified and eliminated. Conduct routine inspection of the elevator pit for water seepage due to faulty waterproofing membrane.
Lift lobbies with poor accessibility for the disabled 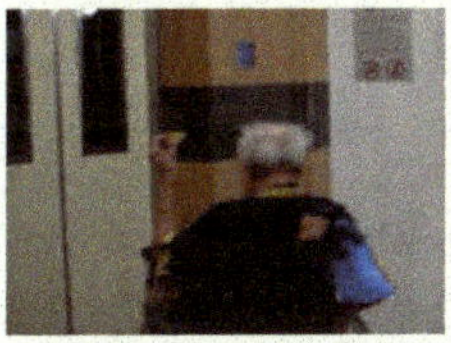 Good practice: Design a wheelchair accessible lift lobby (photo shows an example from a residential building)	Encourage through or two-end entries for lift lobbies; or provide added space for dead-end lobbies to ensure the better distribution of waiting passengers (BS 5655-6:2011, BS 5655-11:2005, BS EN 81-20:2014, SS 550:2009). Provide rain covers for lift lobbies in residential buildings for protection from torrential rains.	Install lift lobby pedestrian flooring as per recommended minimum pendulum ratings specified in SS 485:2011. (See also AS HB 197:1999 and AS/NZS 4663:2004).	Practice proper housekeeping to keep the lift lobbies clean and clear of dirt, and avoid any obstruction or stacking to accommodate easy egress and ingress. Conduct routine inspection of lift call buttons and indicator displays; check that they are in acceptable working condition.

6.5.2 Common Faults

Problem	Design	Construction	Maintenance
Inaccurate elevator car levelling with the landing  Elevator car is not levelled with the landing floor 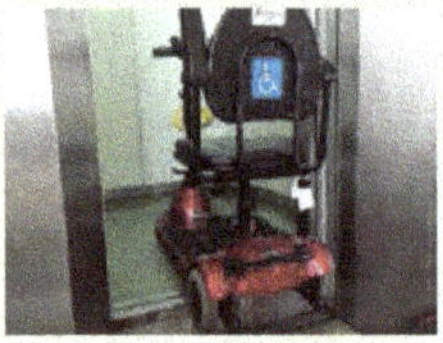 Poor levelling can cause trip hazard	Comply with the design guidelines for permanently installed electric lifts as per SS 550:2009, CIBSE Guide D [14] (BS 5655-6:2011, BS 5655-11:2005, BS EN 81-20:2014) [15]. The stopping accuracy of the elevator car against the landing floor must be ±10mm [13]. Collate global standards on lift safety as per ISO/TR 11071-2:2006 (i.e. assumption of safe operation assured to 125% of rated load, assuring reliability of electric safety devices, mechanical devices built and maintained according to good practice).	Conform to the guidelines for construction and installation of permanently installed electric lifts as per SS 550:2009, BS 5655-6:2011, BS 5655-11:2005, BS EN 81-20:2014. Adhere to the safety rules for construction as per BS EN 81-50:2014. Every lift must be provided with a capacity plate located in a conspicuous place inside the car, indicating the rated load in kilograms and, in the case of passenger lifts, the maximum number of passengers to be carried (BS 5655-6:2011, BS 5655-11:2005, BS EN 81-20:2014, SS 550:2009).	Upgrade the control system, braking, and motor types. A micro-processor controller will electronically monitor and control motor rotation to ensure that the elevator car accurately stops at floor level. Review levelling of car to ensure the value is acceptable by standards to avoid risk of passengers tripping and falling (e.g. wheelchair users) [13]. Permit to operate (PTO) to be displayed in the lifts [13]. Measure ride quality of elevators as per BS ISO 18738:2003.
Faulty door operation  The elevator inner doors failed to close completely during operation (Photo credit: Ben Tng, 2016)	The gap for the elevator car doorway must not exceed 12mm, and the clearance between elevator car door panels must be less than 10mm. The elevator car must not make any movement if the car doors and landing doors are not properly closed and locked [13]. Use durable materials for the doors and more	Test elevator car and landing doors to withstand an impact that is similar to the impact when a person collides with the door at running speed (BS EN 81-20:2014). The main guiding elements of door should operate as intended. Doors must include retainers to keep the door panels in place (BS EN 81-20:2014).	Inspect elevator door and guides, shoes, and tracks. It should not show any permanent deformation or elastic deformation not greater than 15mm (when force of >300 N is applied to 5cm^2 area at centre of door panels at a right angle) (BS 5655-6:2011, BS 5655-11:2005, BS 7255:2012, BS EN 13015:2001+A1:2008, SS 550:2009).

(Continued)

(*Continued*)

Problem	Design	Construction	Maintenance
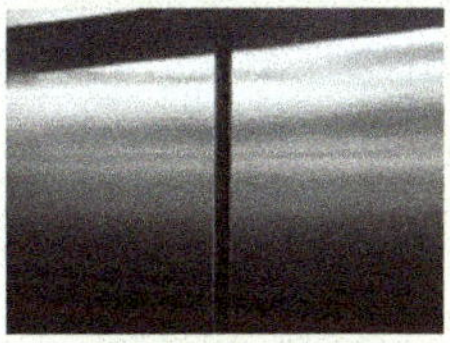 Jammed elevator doors do not close completely	durable materials for their frames (e.g. metals) (BS 5655-6:2011, BS 5655-11:2005, BS EN 81-20:2014, SS 550:2009).	Partially-closed door must open (if button controlling door opening is pressed); while the door must remain open when the door open button is pressed [13].	Review the service call frequency for the door. Increased service calls signify the need to upgrade/replace door operators.

6.5.3 *Elevator Safety*

Problem	Design	Construction	Maintenance
Compromised safety and reliability 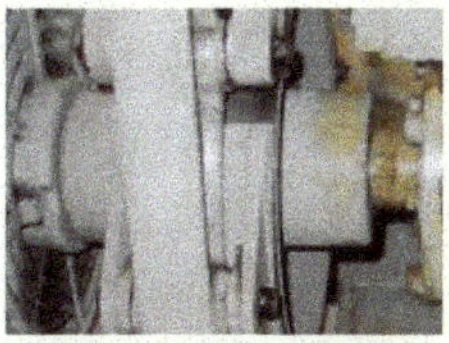 Dirt and rust in machine  Corroded guide rail	Specify electronic components for elevator system reliability (BS 5655-6:2011, BS 5655-11:2005, BS EN 81-20:2014, SS 550: 2009). Ensure compliance to global essential safety requirements (GESRs) for lifts and local safety standards as per ISO/ DTS 8100-21. Lift must be designed to ensure all lift parts do not affect safe operation under reasonable levels of depreciation [13].	Map out safety checks for lifts and classify them according to safety and comfort requirements. Safety gear must be able to stop/ hold lift car and counterweight within allowable distance as per SS 550:2009. (See also BS 5655-6:2011). Proper installation and commissioning of Emergency Battery Operated Power Supply (EBOPS) of lift car, braking system, call buttons, load alarm, safety switches functions, safety logic, emergency lighting and supply, etc. [13].	Conduct conformance test for electronic components of lift machines which are susceptible to damage from high temperatures that may impair reliability (ISO/TR 25743:2010) Conduct monthly safety-levelling of the car and landing. Follow the mandatory incident reporting procedure in the case of an accident/ incident [13].

(*Continued*)

(*Continued*)

Problem	Design	Construction	Maintenance
Faulty suspension ropes due to overloading  Elevator suspension ropes severely damaged  Condition of (broken) ropes in addition to presence of rouge indicate ropes need to be replaced 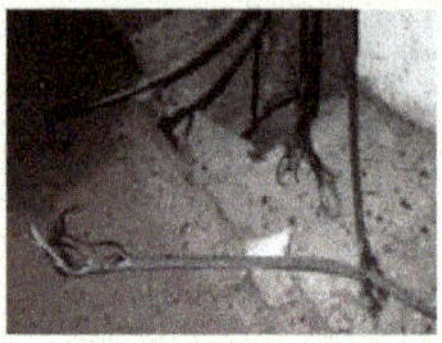 Broken suspension ropes	Comply with the design requirements for rated loading capacity as per SS 550:2009 (BS 5655-6:2011, BS 5655-11:2005, BS EN 81-20:2014). Adhere to the guidelines for lift installation as per ISO/CD 8100-30, ISO 7465:2007, ISO 14798:2009, ISO/NP TR 16765. Provide a lift monitoring system to enable remote tracking of lift breakdown. Consider designing separate service and passenger lifts. Specify durable materials for service lift floor and walls to withstand rough usage.	Overload weighing device should be provided and must activate an alarm when the load in the car exceeds the rated capacity (BS 5655-6:2011, BS 5655-11:2005, BS EN 81-20:2014, SS 550:2009). Install suspension ropes as per ISO 2408:2004, BS EN ISO 16841:2014 and ensure it is properly and equally tensioned. If rope is damaged during installation, even if it passed tests prior to elevator service, the damaged rope should be replaced with a new rope, instead of just replacing the strands (ASME A17.6-2010). Lift capacity rate should be located at a noticeable position in the elevator car. It should indicate the rated load in kilograms and state the maximum number of passengers.	Periodic maintenance should be done by a BCA registered lift contractor at intervals not exceeding one month. An annual inspection and system test should be done by an independent Authorised Examiner (AE). Adhere to the guidelines for the operation and maintenance of permanently installed electric lifts as per SS 550:2009. (See also BS 5655-6:2011, BS 5655-11:2005, BS 7255:2012, BS EN 13015:2001+A1: 2008). Conduct annual test of safety equipment without load. A full load test should be conducted every 5 years. Sufficiently lubricate ropes frequently to avoid abrasive wear between and within the strands. Ensure the timely replacement of ropes if they are permanently kinked, bent, or deformed as per criteria set out in ASME A17.6-2010. Lift ropes shall be tested against requirements of ISO 4344:2004 for signs of excessive wear and tear.

(*Continued*)

(*Continued*)

Problem	Design	Construction	Maintenance
Failure to activate overspeed governor 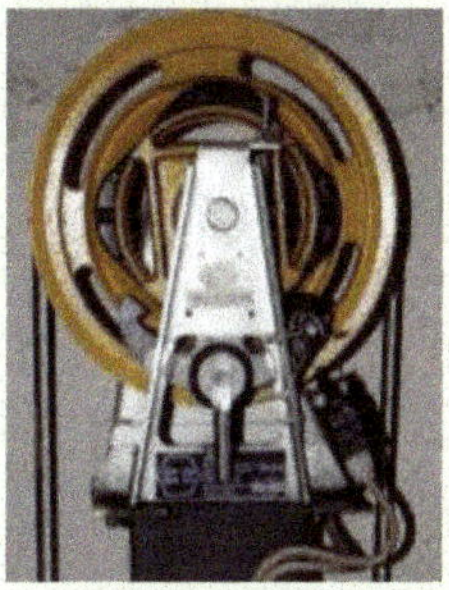 Example of a typical overspeed governor (Photo credit: Jones, I. G. *et al.*, 2013)	The governor ropes should be made from iron, steel, Monel, metal, phosphor bronze or stainless steel (ASME A17.3-2008). The governor rope's diameter should be at least 6mm (BS 5655-6:2011, BS 5655-11:2005, BS EN 81-20:2014, SS 550:2009). The electrical and mechanical tripping speeds should be in accordance with the requirements of SS 550:2009 (BS 5655-6:2011, BS 5655-11:2005, BS EN 81-20:2014).	For safety code for the construction and installation of the overspeed governor refer to BS EN 81-50:2014. For the type of examination for overspeed governors, refer to BS EN 81-50:2014, which requires a minimum of 2 tests conducted with 0,9 — 1,0 gn acceleration to check the strength of the overspeed governor. Comply with BS EN 81-50:2014 for tests of tripping by breakage of suspension means. Commissioning to ensure that overspeed governor functions as intended, for safety under all operating conditions [13].	Inspect the general condition of the speed governor, governor rope and diameter, tripping mechanism and governor switch and governor data plate with no power. Inspect overspeed governor under normal running conditions with power on (BS 5655-6: 2011, BS 5655-11:2005, BS 7255:2012, BS EN 13015:2001+A1:2008, SS 550:2009). Governor rope should not show any sign of excessive wear and tear, in accordance with the requirements in ISO 4344:2004.

6.5.4 *Energy Efficiency*

Problem	Design	Construction	Maintenance
Inefficient energy performance  Good practice: application of LED lighting fittings in an elevator car 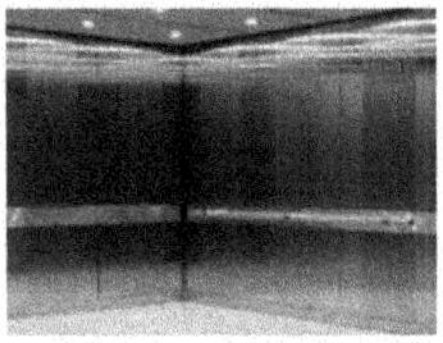 Good practice: use of stainless steel finishing to maximise illumination in car (by reflecting light). Also results in a more durable elevator car	Select and design lift equipment that will cater to expected traffic needs with energy efficiency, as attained by proper equipment management (BS 5655-6:2011, BS 5655-11: 2005, BS EN 81-20:2014, SS 550:2009). Conform to the energy calculation and classification for lifts as per ISO 25745-2:2015. Comply with the minimum energy efficiency requirements as per SS 530:2014. Refer to planning for energy efficiency of lifts and escalators to VDI 4707.	Fulfil energy performance and verification of lifts as per ISO 25745-1:2012. Install luminaires which adhere to the maximum lighting power density for lift lobbies; i.e. 7 W/m^2. Install equipment to measure energy consumption on installed equipment of lifts with reference to ISO 257545 series (Part 1, Part 2 and Part 3). (Refer also to SS 530:2014).	Comply with the measurements of energy consumption. Conform to the energy calculations and classification of escalators and moving walks as per ISO 25745-3:2015. Conduct regular maintenance of the equipment to ensure moving parts are sufficiently lubricated and to identify early signs of wear and tear for timely corrective action and efficient equipment usage. Perform timely modernisation to enhance performance/ energy efficiency (BS 5655-6:2011, BS 5655-11:2005, BS 7255:2012, BS EN 13015:2001+ A1:2008, SS 550:2009).
Poor/compromised lighting  Burnt out elevator car lighting bulb	Lift car should be provided with permanently fixed electric lights (no less than two lighting fittings per car to be provided). Ensure lighting intensity of at least 50 lux at floor level (BS 5655-6:2011, BS 5655-11:2005, BS EN 81-20:2014, SS 550:2009).	Install emergency luminaires in lift cars in accordance with SS 550:2009 (BS 5655-6:2011, BS 5655-11:2005, BS EN 81-20:2014). Use energy efficient lighting with sensors during installation for energy efficiency.	Ensure that luminaires are protected to prevent injury of passengers from breakage; and to prevent access to live parts by passengers (BS 5655-6:2011, BS 5655-11:2005, BS 7255:2012, BS EN 13015:2001+ A1:2008, SS 550:2009).

6.5.5 *Escalators*

Problem	Design	Construction	Maintenance
Escalator and passenger conveyor related maintainability issues  Accumulation of debris within the escalator interior  Footwear getting stuck between steps  Good practice: escalator cordoned off during maintenance operations to prevent unauthorised access	Landing area of escalators and passenger conveyors should have a surface that provides a secure foothold for a minimum distance of 0.85m (measured from the root of the comb teeth). (See also AS 1735.1:2016). Ensure that the escalator and its surroundings have sufficient and adequate illumination. The supporting structure for escalators and passenger conveyors should be designed as per BS EN 115-1:2008+A1:2010, SS 626:2017. Comply with the safety code for elevator design and construction ASME A17.1/CSA B44 — 2010, ISO/DIS 22559-1, BS 8899:2016. Incorporate anti-climbing, anti-sliding, access restriction and deflecting devices to maintain safe operation (SS 626:2017).	Installation of escalators must comply with relevant standards and codes for safety and reliability (BS EN 115-1:2008+A1:2010, SS 626:2017). All machinery must be mounted securely and be defect free (e.g. should not have any oil leakage). To ensure safe operation without issues due to corrosion and wear and tear, all escalator components should be of durable and reliable make. All installed signs, inscriptions and notices should be made of durable materials (BS EN 115-1:2008+A1:2010, SS 626:2017).	Ensure proper housekeeping of escalator to keep it clean and free of debris. Building owner/operator need to conduct monthly maintenance of escalators (including maintenance of safety switches, sensors, emergency stops, and handrails) as per SS 626:2017. (See also BS EN 115-1:2008+A1:2010). The annual inspection and testing should be performed by an independent Authorised Examiner (AE). Access to the escalator or passenger conveyor should be barred by suitable devices and notices/signage displaying "No access/no entry" should be provided during maintenance, repair works, or inspections (BS EN 115-1:2008+A1:2010, SS 626:2017). Adhere to the inspection criteria for safety of escalators as per JIS A 4302:2006. Refer to procedure for ride quality measurements of escalators and moving walks as per BS ISO 18738-2:2003.

References

[1] Wujek, J.B. and Dagostino, F. (2010). *Mechanical and Electrical Systems in Architecture, Engineering, and Construction* (5th ed.). Ohio: Prentice Hall.

[2] ASHRAE (2016). ASHRAE Standard 15 — Safety Standard for Refrigeration Systems and Designation and Classification of Refrigerants. USA: ASHRAE.

[3] Chew, M. Y. L. (2016). *Maintainability of Facilities: Green FM for Building Professionals* (2nd ed.). Singapore: World Scientific.

[4] International Code Council, Inc. (2011). *International Plumbing Code* (2012 ed.). USA: International Code Council, Inc.

[5] Public Utilities Board (2004). *Code of Practice on Sewerage and Sanitary Works* (1st Adden.). Singapore: Public Utilities Board (PUB).

[6] Singapore Civil Defence Force (2015). Singapore Fire Safety Engineering Guidelines. Retrieved on March 22, 2017 from https://www.scdf.gov.sg/sites/www.scdf.gov.sg/files/Singapore%20Fire%20Safety%20Engineering%20Guidelines%202015_1.pdf.

[7] National Fire Protection Association (2016). *NFPA 70B: Recommended Practice for Electrical Equipment Maintenance*. National Fire Protection Association.

[8] SPRING Singapore (1998). *CP 5: Code of practice for electrical installations*. Singapore: SPRING Singapore.

[9] Singapore Civil Defence Force (2013). *Code of Practice for Fire Precautions in Buildings*. Singapore: SCDF.

[10] NUS Maintainability of Buildings (2016). HVAC. Retrieved on January 22 from http://www.hpbc.bdg.nus.edu.sg/?page_id=94&page=4.

[11] Institute of Environmental Epidemiology (2001). *Code of Practice for the Control of Legionella Bacteria in Cooling Towers* (4th ed.). Singapore: Ministry of the Environment.

[12] National Environment Agency (2008). *Code of Practice on Piped Drinking Water Sampling and Safety Plans*. Singapore: National Environment Agency (NEA).

[13] BMSMA (2016). Building Maintenance and Strata Management Act. (Lift, Escalator and Building Maintenance) Regulations 2016.

[14] CIBSE (2015). *CIBSE Guide D: Transportation Systems in Buildings*. UK: CIBSE.

[15] Piper, J. (2004). *Handbook of Facility Assessment*. United States of America: The Fairmount Press.

Normative References/Standards Referred to for Mechanical and Electrical Systems

- AHRI 260 — Sound rating of ducted air moving and conditioning equipment
- ANSI/ASHRAE 188-2015 — Legionellosis: Risk Management for Building Water Systems
- ANSI/ASHRAE/ACCA Standard 180-2012 — Standard practice for inspection and maintenance of com-mercial building HVAC systems
- ANSI/ASHRAE/IES Standard 90.1-2016 — Energy Standard for Buildings Except Low-Rise Residential Buildings

- ANSI/ASHRAE/IES/USGBC Standard 189.1-2014 — Standard for the Design of High-Performance Green Buildings
- AS 1668.2-2012 — The use of ventilation and airconditioning in buildings — Mechanical ventilation in buildings
- AS 1735.1:2016 — Lifts, escalators and moving walks. General requirements
- AS 2293 SET:2005 — Emergency escape lighting and exit signs Set
- AS HB 197:1999 — An introductory guide to the slip resistance of pedestrian surface materials
- AS/NZS 2293.2:1995/AMDT 3:2012 — Emergency escape lighting and exit signs for buildings Inspection and maintenance
- AS/NZS 4663:2004 — Slip resistance measurement of existing pedestrian surfaces
- ASHRAE Guideline 12-2000 — Minimizing the Risk of Legionellosis Associated with Building Water Sys-tems
- ASME A17.1/CSA B44:2010 — Safety Code for Elevators and Escalators
- ASME A17.3:2008 — Safety Code for Existing Elevators and Escalators
- ASME A17.6:2010 — Standard for Elevator Suspension, Compensation, and Governor Systems
- BS 1363-4:2016 — 13 A plugs, socket-outlets, adaptors and connection units. Specification for 13 A fused connection units switched and unswitched
- BS 5266-1:2016 — Emergency lighting. Code of practice for the emergency lighting of premises
- BS 5306-3:2009 — Fire extinguishing installations and equipment on premises. Commissioning and maintenance of portable fire extinguishers. Code of practice
- BS 5306-8:2012 — Fire extinguishing installations and equipment on premises. Selection and positioning of portable fire extinguishers. Code of practice
- BS 5306-9:2015 — Fire extinguishing installations and equipment on premises. Recharging of portable fire extinguishers. Code of practice
- BS 5655-11:2005 — Lifts and service lifts. Code of practice for the undertaking of modifications to exist-ing electric lifts
- BS 5655-6:2011 — Lifts and service lifts. Code of practice for the selection, installation and location of new lifts
- BS 5839-1:2017 — Fire detection and fire alarm systems for buildings. Code of practice for design, in-stallation, commissioning and maintenance of systems in non-domestic premises
- BS 5839-3:1988 — Fire detection and alarm systems for buildings. Specification for automatic release mechanisms for certain fire protection equipment
- BS 5839-6:2013 — Fire detection and fire alarm systems for buildings. Code of practice for the design, installation, commissioning and maintenance of fire detection and fire alarm systems in domestic prem-ises
- BS 5839-9:2011 — Fire detection and fire alarm systems for buildings. Code of practice for the design, installation, commissioning and maintenance of emergency voice communication systems

- BS 5908-1:2012 — Fire and explosion precautions at premises handling flammable gases, liquids and dusts. Code of practice for precautions against fire and explosion in chemical plants, chemical storage and similar premises
- BS 6391:2009 — Specification for non-percolating lay flat delivery hoses and hose assemblies for fire-fighting purposes
- BS 6423:2014 — Code of practice for maintenance of low-voltage switchgear and control gear
- BS 6626:2010 — Maintenance of electrical switchgear and control gear for voltages above 1 kV and up to and including 36 kV. Code of practice
- BS 7255:2012 — Code of practice for safe working on lifts
- BS 7291-1:2010 — Thermoplastics pipe and fitting systems for hot and cold water for domestic purpos-es and heating installations in buildings. General requirements
- BS 7430:2011+A1:2015 — Code of practice for protective earthing of electrical installations
- BS 7671:2008+A3:2015 — Requirements for Electrical Installations. IET Wiring Regulations
- BS 7698-7:1996, ISO 8528-7:1994 — Reciprocating internal combustion engine driven alternating current generating sets. Technical declarations for specification and design
- BS 8486-1:2007+A1:2011 — Examination and test of new lifts before putting into service. Specification for means of determining compliance with BS EN 81. Electric lifts
- BS 8512:2008 — Electric cables. Code of practice for the storage, handling, installation and disposal of cables on wooden drums
- BS 8554:2015 — Code of practice for the sampling and monitoring of hot and cold water services in buildings
- BS 8558:2015 — Guide to the design, installation, testing and maintenance of services supplying water for domestic use within buildings and their curtilages. Complementary guidance to BS EN 806
- BS 8899:2016 — Improvement of fire-fighting and evacuation provisions in existing lifts. Code of prac-tice
- BS 9990:2015 — Non automatic fire-fighting systems in buildings. Code of practice
- BS EN 1004:2004 — Mobile access and working towers made of prefabricated elements. Materials, di-mensions, design loads, safety and performance requirements
- BS EN 10088-2:2014 — Stainless steels. Technical delivery conditions for sheet/plate and strip of corro-sion resisting steels for general purposes
- BS EN 1057:2006+A1:2010 — Copper and copper alloys. Seamless, round copper tubes for water and gas in sanitary and heating applications
- BS EN 115-1:2008+A1:2010 — Safety of escalators and moving walks. Construction and installation
- BS EN 13015:2001+A1:2008 — Maintenance for lifts and escalators. Rules for maintenance instructions
- BS EN 13121-3:2016 — GRP tanks and vessels for use above ground. Design and workmanship

- BS EN 1402:2009 — Rubber and plastics hoses and hose assemblies. Hydrostatic testing
- BS EN 1567:1999 — Building valves. Water pressure reducing valves and combination water reducing valves. Requirements and tests.
- BS EN 16767:2016 — Industrial valves. Steel and cast iron check valves
- BS EN 1796:2013 — Plastics piping systems for water supply with or without pressure. Glass-reinforced thermosetting plastics (GRP) based on unsaturated polyester resin (UP)
- BS EN 1838:2013 — Lighting applications. Emergency lighting
- BS EN 1947:2014 — Fire-fighting hoses. Semi-rigid delivery hoses and hose assemblies for pumps and vehicles
- BS EN 1982:2008 — Copper and copper alloys. Ingots and castings
- BS EN 1992-1-1:2004+A1:2014 — Eurocode 2: Design of concrete structures. General rules and rules for buildings
- BS EN 3 series — Portable fire extinguishers
- BS EN 50172:2004, BS 5266-8:2004 — Emergency escape lighting systems
- BS EN 545:2010 — Ductile iron pipes, fittings, accessories and their joints for water pipe-lines. Require-ments and test methods
- BS EN 598:2007+A1:2009 — Ductile iron pipes, fittings, accessories and their joints for sewerage appli-cations. Requirements and test methods
- BS EN 61009-2-1:1995 — Specification for residual current operated circuit-breakers with integral overcur-rent protection for household and similar uses (RCBOs). Applicability of the general rules to RCBOs functionally independent of line voltage
- BS EN 62305-1:2011 — Protection against lightning. General principles
- BS EN 694:2014 — Fire-fighting hoses. Semi-rigid hoses for fixed systems
- BS EN 805:2000 — Water supply. Requirements for systems and components outside buildings
- BS EN 806-5:2012 — Specifications for installations inside buildings conveying water for human con-sumption. Operation and maintenance
- BS EN 81-20:2014 — Safety rules for the construction and installation of lifts. Lifts for the transport of persons and goods. Passenger and goods passenger lifts
- BS EN 81-50:2014 — Safety rules for the construction and installation of lifts. Examinations and tests. Design rules, calculations, examinations and tests of lift components
- BS EN ISO 16841:2014 — Steel wire ropes. Pulling eyes for rope installation. Types and minimum re-quirements
- BS EN ISO 21003-2:2008+A1:2011 — Multilayer piping systems for hot and cold water installations in-side buildings. Pipes
- BS EN ISO 21003-3:2008 — Multilayer piping systems for hot and cold water installations inside build-ings. Fittings
- BS EN ISO 21003-5:2008 — Multilayer piping systems for hot and cold water installations inside build-ings. Fitness for purpose of the system
- BS ISO 10916:2014 — Calculation of the impact of daylight utilization on the net and final energy de-mand for lighting
- BS ISO 18738:2003 — Lifts (elevators). Measurement of lift ride quality

- BS ISO 18738-2:2012 — Measurement of ride quality — Part 2: Escalators and moving walks
- CSA Z412-2000 (R2016) — Guideline on Office Ergonomics
- IEC 60079 — Explosive Atmosphere Standards
- IEC 61009-1:2010+AMD1:2012+AMD2:2013 CSV (Consolidated version) — Residual current operated circuit-breakers with integral overcurrent protection for household and similar uses (RCBOs) — Part 1: General rules
- IEC 62305-3:2010 — Protection against lightning — Part 3: Physical damage to structures and life hazard
- IEC 62305-4:2010 — Protection against lightning — Part 4: Electrical and electronic systems within struc-tures
- ISO 10816-3:2009 — Mechanical vibration — Evaluation of machine vibration by measurements on non-rotating parts — Part 3: Industrial machines with nominal power above 15 kW and nominal speeds be-tween 120 r/min and 15 000 r/min when measured in situ
- ISO 12242:2012 — Measurement of fluid flow in closed conduits — Ultrasonic transit-time meters for liq-uid
- ISO 13612-1:2014 — Heating and cooling systems in buildings — Method for calculation of the system performance and system design for heat pump systems — Part 1: Design and dimensioning
- ISO 1452-1:2009 — Plastics piping systems for water supply and for buried and above-ground drainage and sewerage under pressure — Unplasticized poly(vinyl chloride) (PVC-U) — Part 1: General
- ISO 14798:2009 — Lifts (elevators), escalators and moving walks — Risk assessment and reduction methodology
- ISO 16814:2008 — Building environment design — Indoor air quality — Methods of expressing the quali-ty of indoor air for human occupancy
- ISO 2017-1:2005 — Mechanical vibration and shock — Resilient mounting systems — Part 1: Technical information to be exchanged for the application of isolation systems
- ISO 2230:2002 — Rubber products — Guidelines for storage
- ISO 2408:2004 — Steel wire ropes for general purposes — Minimum requirements
- ISO 25745-1:2012 — Energy performance of lifts, escalators and moving walks — Part 1: Energy meas-urement and verification
- ISO 25745-2:2015 — Energy performance of lifts, escalators and moving walks — Part 2: Energy calcula-tion and classification for lifts (elevators)
- ISO 25745-3:2015 — Energy performance of lifts, escalators and moving walks — Part 3: Energy calcula-tion and classification of escalators and moving walks
- ISO 29463-1:2011 — High-efficiency filters and filter media for removing particles in air — Part 1: Classi-fication, performance testing and marking
- ISO 29463-5:2011 — High-efficiency filters and filter media for removing particles in air — Part 5: Test method for filter elements
- ISO 3864-1:2011 — Graphical symbols — Safety colours and safety signs — Part 1: Design principles for safety signs and safety markings
- ISO 4344:2004 — Steel wire ropes for lifts — Minimum requirements

- ISO 5149-1:2014/Amd 1:2015 — Refrigerating systems and heat pumps — Safety and environmental requirements — Part 1: Definitions, classification and selection criteria AMENDMENT 1: Correction of QLAV, QLMV
- ISO 6182 series — Fire protection — Automatic sprinkler systems
- ISO 7240 series — Fire detection and alarm systems
- ISO 7465:2007 — Passenger lifts and service lifts — Guide rails for lift cars and counter-weights — T-type
- ISO 8995-1:2002/Cor 1:2005 — Lighting of work places — Part 1: Indoor
- ISO 9996:1996 — Mechanical vibration and shock — Disturbance to human activity and performance — Classification
- ISO/CD 8100-30 — Lifts for the transport of persons and goods — Part 30: Class I, II, III and VI lifts in-stallation
- ISO/DIS 22559-1 — Safety requirements for lifts (elevators) — Part 1: Global essential safety require-ments (GESRs)
- ISO/DTS 8100-21 — Lifts for the transport of persons and goods — Part 21: Global safety parameters (GSPs) meeting the global essential safety requirements (GESRs)
- ISO/FDIS 13253 — Ducted air-conditioners and air-to-air heat pumps — Testing and rating for perfor-mance
- ISO/NP TR 16765 — Comparison of worldwide safety standards on lifts for firefighters
- ISO/PRF 7165 — Firefighting — Portable fire extinguishers — Performance and construction
- ISO/TR 11071-2:2006 — Comparison of worldwide lift safety standards — Part 2: Hydraulic lifts (eleva-tors)
- ISO/TR 25743:2010 — Lifts (elevators) — Study of the use of lifts for evacuation during an emergency
- JIS A 4302:2006 — Inspection standard of elevator, escalator and dumbwaiter
- JIS A 4422:2011 — Toilet seat with shower unit
- NFPA 10:2013 — Standard for Portable Fire Extinguishers
- NFPA 101 — Life Safety Code
- NFPA 110:2016 — Standard for Emergency and Standby Power Systems
- NFPA 13:2016 — Standard for the Installation of Sprinkler Systems
- NFPA 14:2016 — Standard for the Installation of Standpipe and Hose Systems
- NFPA 25:2017 — Standard for the Inspection, Testing, and Maintenance of Water-Based Fire Protection Systems
- NFPA 72:2016 — National Fire Alarm and Signaling Code
- NFPA 780 — Standard for the Installation of Lightning Protection Systems
- NFPA 80 — Standard for Fire Doors and Other Opening Protectives
- SS 141:2013 — Specification for unplasticised PVC pipe for cold water services and industrial uses
- SS 245:2014 — Specification for glass reinforced polyester sectional water tanks
- SS 332:2007 — Specification for fire doors
- SS 375-1:2015 — Suitability of non-metallic products for use in contact with water intended for human consumption with regard to their effect on the quality of the water — Part 1 : Specification

- SS 403:2013 — Specification for 13A fused connection units switched and unswitched
- SS 480:2016 — Residual current operated circuit-breakers with integral overcurrent protection for house-hold and similar uses (RCBOs) — General rules
- SS 485:2011 — Specification for slip resistance classification of pedestrian surface materials
- SS 508-3:2013 — Graphical symbols — Safety colours and safety signs — Design principles for graph-ical symbols for use in safety signs
- SS 514:2016 — Code of practice for office ergonomics
- SS 530:2014 — Code of practice for energy efficiency standard for building services and equipment
- SS 531-1:2006(2013) — Code of practice for lighting of work places — Indoor
- SS 532:2016 — Code of practice for the storage of flammable liquids
- SS 535:2007 — Code of practice for installation, operation, maintenance, performance and construction requirements of mains failure standby generating systems
- SS 538:2008 — Code of practice for maintenance of electrical equipment of electrical installations
- SS 546:2009 — Code of practice for emergency voice communication systems in buildings
- SS 550:2009 — Code of practice for installation, operation and maintenance of electric passenger and goods lifts
- SS 551:2009 — Code of practice for earthing
- SS 553:2016 — Code of practice for air-conditioning and mechanical ventilation in buildings
- SS 554:2016 — Code of practice for indoor air quality for air-conditioned buildings
- SS 555-1:2010 — Code of practice for protection against lightning — Part 1: General principles
- SS 555-3:2010 — Code of practice for protection against lightning — Part 3: Physical damage to struc-tures and life hazard
- SS 555-4:2010 — Code of practice for protection against lightning — Part 4: Electrical and electronic systems within structures
- SS 563-1:2010 — Code of practice for the design, installation and maintenance of emergency lighting and power supply systems in buildings — Part 1: Emergency lighting
- SS 563-2:2010 — Code of practice for the design, installation and maintenance of emergency lighting and power supply systems in buildings — Part 2: Installation requirements and maintenance procedures
- SS 564-1:2013 — Green data centres — Part 1: Energy and environmental management systems
- SS 564-2:2013 — Singapore Standard for green data centres — Part 2: Guidance for energy and envi-ronmental management systems
- SS 575:2012 — Code of practice for fire hydrant, rising mains and hose reel systems
- SS 578:2012 — Code of practice for use and maintenance of portable fire extinguishers
- SS 591:2013 — Code of practice for long term measurement of central chilled water system energy effi-ciency
- SS 626:2017 — Code of practice for design, installation and maintenance of escalators and moving walks

- SS CP 10:2005 — Code of practice for the installation and servicing of electrical fire alarm systems
- SS CP 48:2005 — Code of practice for water services
- SS CP 5:1998 — Code of practice for electrical installations
- SS CP 52:2004 — Code of practice for automatic fire sprinkler system
- SS CP 82:1999 — Code of practice for waterproofing of reinforced concrete buildings
- SS CP 99:2003 — Code of practice for industrial noise control
- SS EN 3 Series (SS EN 3–7:2012) — Portable fire extinguishers — Characteristics, performance require-ments and test methods
- SS EN 3-7:2012 — Portable fire extinguishers — Characteristics, performance requirements and test methods
- SS IEC 60598-1:2016 — Luminaires — Part 1: General requirements and tests
- SS ISO 22301:2012 — Societal security — Business continuity management systems — Requirements
- SS ISO 22313:2013 — Business continuity management systems — Guidance
- VDI 4707 — Lifts Energy Efficiency

Index

www.ingramcontent.com/pod-product-compliance
Ingram Content Group UK Ltd.
Pitfield, Milton Keynes, MK11 3LW, UK
UKHW051044150926
8624IPUK00006B/387

9 789813 230590